PHYSICS PROJECT LAB

Physics Project Lab

Paul Gluck
Jerusalem College of Engineering

John King
Massachusetts Institute of Technology

OXFORD
UNIVERSITY PRESS

Great Clarendon Street, Oxford, OX2 6DP,
United Kingdom

Oxford University Press is a department of the University of Oxford.
It furthers the University's objective of excellence in research, scholarship,
and education by publishing worldwide. Oxford is a registered trade mark of
Oxford University Press in the UK and in certain other countries

© Paul Gluck and John King 2015

The moral rights of the authors have been asserted

First Edition published in 2015

Impression: 1

All rights reserved. No part of this publication may be reproduced, stored in
a retrieval system, or transmitted, in any form or by any means, without the
prior permission in writing of Oxford University Press, or as expressly permitted
by law, by licence or under terms agreed with the appropriate reprographics
rights organization. Enquiries concerning reproduction outside the scope of the
above should be sent to the Rights Department, Oxford University Press, at the
address above

You must not circulate this work in any other form
and you must impose this same condition on any acquirer

Published in the United States of America by Oxford University Press
198 Madison Avenue, New York, NY 10016, United States of America

British Library Cataloguing in Publication Data

Data available

Library of Congress Control Number: 2014940248

ISBN 978–0–19–870457–7 (hbk.)
ISBN 978–0–19–870458–4 (pbk.)

Printed and bound by
CPI Group (UK) Ltd, Croydon, CR0 4YY

Links to third party websites are provided by Oxford in good faith and
for information only. Oxford disclaims any responsibility for the materials
contained in any third party website referenced in this work.

This book is dedicated to the memory of my co-author Professor John King, a great physicist and educator and a fine, warm, human being.

Paul Gluck

Acknowledgments

We are very grateful to Professors Charles Holbrow and Gabriel Spalding for their critical reading of the manuscript and making extremely useful suggestions which we adopted in the revised version.

Special thanks are due to Jessica White, Victoria Mortimer, and Richard Hutchinson at Oxford University Press for all the help and kindness extended to us throughout the process of publication.

Contents

Figure acknowledgments   x

Projects: why and how?   1

Part 1 Mechanics

1. Bouncing balls   13
2. Mechanics of soft springs   17
3. Pulse speed in falling dominoes   25
4. A variable mass oscillator   28
5. Rotating vertical chain   34
6. Cycloidal paths   38
7. Physics of rubber bands and cords   44
8. Oscillation modes of a rod   49

Part 2 Electromagnetism

9. Physics of incandescent lamps   57
10. Propulsion with a solenoid   62
11. Magnetic dipoles   69
12. The jumping ring of Elihu Thomson   74
13. Microwaves in dielectrics I   80
14. Microwaves in dielectrics II   86
15. The Doppler effect   89
16. Noise   94
17. Johnson noise   102
18. Network analogue for lattice dynamics   106
19. Resistance networks   111

Part 3 Acoustics

20 Vibrating wires and strings — 117
21 Physics with loudspeakers — 124
22 Physics of the tuning fork — 129
23 Acoustic resonance in pipes — 134
24 Acoustic cavity resonators and filters — 138
25 Room acoustics — 141
26 Musical instruments: the violin — 146
27 Musical instruments: the guitar — 151
28 Brass musical instruments — 155

Part 4 Liquids

29 Sound from gas bubbles in a liquid — 163
30 Shape and path of air bubbles in a liquid — 168
31 Ink diffusion in water — 173
32 Refractive index gradients — 176
33 Light scattering by surface ripples — 180
34 Diffraction of light by ultrasonic waves in liquids — 184
35 The circular hydraulic jump — 188
36 Vortex physics — 192
37 Plastic bottle oscillator — 197
38 Salt water oscillator — 201

Part 5 Optics

39 Birefringence in cellulose tapes — 207
40 Barrier penetration — 212
41 Reflection and transmission of light — 215
42 Polarization by transmission — 221
43 Laser speckle — 226

44	Light scattering from suspensions	232
45	Light intensity from a line source	236
46	Light interference in reflecting tubes	239

Part 6 Temperature and Heat

47	Cooling I	245
48	Cooling II	248
49	The Leidenfrost effect I	254
50	The Leidenfrost effect II: drop oscillations	258
51	The drinking bird	260
52	Liquid–vapor equilibrium	266
53	Solar radiation flux	270

Appendix A:	Project ideas	273
Appendix B:	Facilities, materials, devices, and instruments	290
Appendix C:	Reference library	310

Index	315

Figure acknowledgments

Explanation of figure acknowledgments:

4.4, 4.5: **71**, 721(2003) means Figures 4.4 and 4.5 in the text, followed by the journal reference (volume number, page number, publication year).

Figures in the text reproduced with permission from journals published by the American Institute of Physics, by permission of the American Association of Physics Teachers:

American Journal of Physics:
4.4, 4.5: **71**, 721(2003); 11.4: **62**, 702(1994); 13.2: **48**, 648(1980); 13.4: **54**, 712(1986); 13.5: **52**, 214(1984); 14.2: **45**, 88(1977); 20.2: **53**, 479(1985); 35.3: **67**, 723(1999); 36.2, 36.4: **75**, 1092(2007); 37.5: **75**, 893(2007); 39.4: **41**, 1184(1973); 40.3: **43**, 107(1975); 41.4: **50**, 158(1982); 45.2: **63**, 47(1995); 48.2: **61**, 568(1993); 52.2: **64**, 1165(1996).

The Physics Teacher:
9.2: **45**, 466(2008); 23.2, 23.3: **44**, 10(2006).

Physics of Fluids A:
30.2: **14**, L49(2002).

Figures in the text reproduced from journals published by the Institute of Physics UK, by permission of IOP Publishing:

Physics Education:
1.4: **33**, 174(1998); 1.5: **33**, 236(1998); 2.9: **45**, 178(2010); 6.6–6.11: **45**, 176(2010); 10.3, 10.4, 10.9: **45**, 466(2008); 12.7: **7**, 238(1972); 28.1, 28.2, 28.3: **38**, 300(2003).

European Journal of Physics:
18.7: **1**, 129(1980); 29.3, 29.4: **8**, 98(1987); 21, 571(2000); 44.3: **7**, 259(1986); 46.3: **13**, 47(1992); 48.3, 48.4, 48.5: **30**, 559(2009).

Projects: why and how?

Introduction

This book is intended for our colleagues, teachers of physics at universities and colleges, and their students. While significant reference is made to material typically encountered *beyond the first year* of university-level study, high school teachers and their students may also find many opportunities here. Like us, many readers have probably been teaching physics most of their lives in what may be called "transmission" mode, that is, instruction by means of textbook–lesson–recitation–laboratory. In this mode we, the teachers, send, and our charges, the students, receive information and skills, in conformity with John Locke's notion that the student's mind resembles an empty slate to be written on.

But there is another mode of instruction, exemplified by what Rousseau wrote in his book *Emile*: "Teach by doing whenever you can, and only fall back on words when doing is out of the question." We may call this the transformation mode. Now, both teacher and student are transformed into explorers, with benefits to both. For the teachers in a research-based institution nothing should be more natural than to carry over the research style into some of their teaching activity. Why this does not happen more often deserves serious discussion. For teachers in a high school it is an opportunity to partake in and guide research, and this could prove to be the activity that relieves repetition and lifts them from the rut. For the student it is a chance to experience the way science works.

An important aspect of the project lab is that the teacher, though generally experienced, doesn't always know the answers to the questions raised in the project any more than the students do. So much so that instead of the words *teacher* and *student* we will use *guide* and *researcher*. (Except when we quote.)

The two modes of learning should be combined in a proportion that suits the personalities of guides and researchers alike. It is not our aim in this book to evaluate the competing modes. The research literature in physics teaching in the past 30 years has been devoted to that, and it behooves us to be mindful of it. Rather, we should like to extend an invitation to the reader to try something different, at least as far as laboratory work is concerned, namely, to teach physics by guiding researchers through projects. It may happen that once tried, the experience will be habit forming, and that shall be our reward for writing this book. For those of our readers who have guided projects for years these remarks are like preaching to the converted. We hope that for them this book will be a source of further ideas.

Introduction	1
What is a project?	2
Overcoming the fear	4
Tools of the trade	5
Researchers, choices, and the toil	7
Is it worth the effort?	9
About the book	9

What is a project?

Greg Lockett, a teacher who participated in the LabNet project has described important aspects of project lab [1]:

> The goal of the guide is to re-create the process and experience of working in the physics laboratory. Several features are important in this approach. Students are free to choose a research problem. They do not know the solution to their problem at the outset. While they work within the constraint of time, resources and their current knowledge, they are given sufficient time and freedom to attack their problem as they see fit. Collaboration is acceptable and desirable. The end product can vary: reports, papers, equipment, experiments, models, *etc*.
>
> In this process the goal is to give the student an immediate "inside" view of how physics is done. The teacher functions as a facilitator, somewhat like a research director in a laboratory: obtains needed resources or suggests alternatives, resolves disputes about equipment scheduling and squabbles between partners, attempts to foster new ideas and techniques, and strengthens theory when needed.
>
> Skills are learned as needed. These include guided journal, book and Internet searches, motor skills in building apparatus, social skills of collaboration, communication skills (oral, writing, optimal presentation), and mathematical and analytical skills. The project will demand these naturally, so problems of motivation will not arise.

We would also like to quote two people involved in experimental physics who describe what we feel ought to be the approach to doing projects; namely, that the researcher should acquire the beginnings of various skills that were once essential in experimental physics. Although in recent decades digital instruments have largely supplanted analog techniques, their inner workings are often not well understood by the user. In spite of the value and importance of digital techniques, we think that beginner researchers would also benefit from hands-on experience with earlier, understandable analog techniques.

Here is what Nobel Prize winner P.M.S. Blackett had to say [2]:

> The experimental physicist is a jack-of-all-trades, a versatile but amateur craftsman. He must blow glass and turn metal, though he could neither earn his living as a glass-blower nor be classed as a skilled mechanic; he must do carpentry, photograph, wire electric circuits and be a master of gadgets of all kinds; he may find invaluable a training as an engineer and can always profit by utilizing his gifts as a mathematician. In such activities he will be engaged for three-quarters of his working day. During the rest he must be a physicist, that is, he must cultivate an intimacy

with the behaviour of the physical world. Bur in none of these activities, taken alone, need he be pre-eminent, certainly neither as a craftsman, for he will seldom achieve more than an amateur's skill, nor even in his knowledge of his own special field of physics need he, or indeed perhaps can he, surpass the knowledge of some theoretician. For a theoretical physicist has no long laboratory hours to keep him from study, and he must in general be accredited with at least an equal physical intuition and certainly a greater mathematical skill. The experimental physicist must be enough of a theorist to know what experiments are worth doing and enough of a craftsman to be able to do them. He is only pre-eminent in being able to do both.

In a more general vein Frank O'Brien, a senior research technician in the MIT molecular beam laboratory, and a guide over many years to more than 150 graduate and undergraduate researchers, has this to say [3]:

Listen to honest advice with an open mind, have an ear that listens *and* comprehends. Practice ingenuity with simplicity. Be patient and comprehend what you observe. Know how to relax. Have a sense of humor. Have some ability to relate to people. Train to have a steady hand. Be reasonably versed in sketching, drafting, machining, measuring accurately when needed, plumbing, circuit wiring, jewelry work, soldering, brazing, welding and trouble-shooting.

A project is the experimental exploration of physical phenomena, coupled with some understanding of the theory behind them. On the one hand, this definition is flexible enough to subsume the behavior of everyday objects and phenomena, and to cater for the curiosity and enthusiasm of any researcher or guide interested in them. On the other hand, it is a commitment to measure, collect, classify, and analyze data in a controlled manner, and if possible understand it in terms of some model or theory, and not to be satisfied with building gadgets or apparatus, however ingenious.

Now where does one get ideas for projects? For the curious and the observant they are everywhere. Seemingly trivial occurrences can be investigated thoroughly, at increasing depths of sophistication, and may engage the attention of researcher and guide alike for time spans that go way beyond the initial estimate. We hope to demonstrate this for a number of cases in later chapters. Sometimes one can pair researchers with a common interest, as deduced from a simple questionnaire; for instance, two string instrument players who would be interested in how a bow excites and damps a string, or the acoustical effect of varnish on the wood. The underlying interest and enthusiasm in these cases is invaluable. Here are a couple of examples, described in later chapters, to whet the appetite and to calm readers who might panic, not knowing what subjects to offer to or accept from their students.

What regularities are there in the time-dependence of the "glug-glug" emptying of a bottle turned upside down? What if two such bottles are connected: will they empty "in phase?"

Water from a tap hits a sink and a circle of liquid is formed there. What factors influence the radius of this circle?

Indeed, the behavior of any object under changing circumstances, or any natural phenomenon, reveals physical properties which can be measured, their correlations explored, their behavior understood and therefore often controlled. There is usually more than meets the eye, and the uncovering of that is the stuff of which projects are made.

Overcoming the fear

It can be a long path from defining a project to knowing how to navigate it to a reasonably satisfying conclusion. Various problems with instrumentation, construction, getting data, mathematical analysis, etc., may arise, along with frustrations and dead ends, all coped with by helpful and continual guide–researcher interaction. Inevitably there is an element of uncertainty and adventure in all this, especially if the topic does not appear in a book, and has no standard answer known in advance.

A certain confidence is required in the process, in which the guide sometimes has only the advantage of experience over the researcher. Confidence can only be acquired by plunging in, not perhaps at the deep end but in the middle. By this we mean adapting existing experiments and investigations to the school's milieu. We hope that the material presented in this book will be helpful in this respect. The process may be likened to the comedian who begins by telling the jokes of others, or those scripted for him or her. As expertise and confidence grows, and the tools of the trade are acquired, the stage of stand-up comedy is reached, in varying degrees of dexterity.

There is a large body of experimental papers in the teaching literature, in journals like the *American Journal of Physics*, *Physics Education*, and the *European Journal of Physics*. Indeed, many of the projects described here have their origin in adaptation of articles in these journals. There is enough of a spectrum here, both in subjects and in level, to appeal to a wide range of people wishing to start. For that very reason the levels of the projects described in this book are uneven. The website of the International Young Physicists Tournament, <http://www.iypt.org>, is also a good source for projects.

Some of the tools to be acquired will be discussed in the following sections. These, in combination with experience, will lead to an important, yet somewhat intangible, expertise: one of knowing one's limitations, what is doable and understandable with one's knowledge, equipment, budget, and time limitations.

A colleague of ours has been doing projects for years. He has several times opined that he will not suggest a topic to a researcher unless he knows in advance how to solve it all the way. Of course this "guarantees" success. Nevertheless, we do not insist on adhering to this amount of certainty. It almost defeats the purpose, and diminishes the mystery and fun.

Tools of the trade

The scope and the sophistication of a project will naturally depend on expertise, equipment, manpower, and the ability and willingness to learn. Sealing wax and string cannot go very far. Experience has taught us to look out for the following.

Shaping materials and signals

Knowing where to get the services of a machine shop and a carpenter is very useful. You shape metal, plastics, glass, and wood to make apparatus perform functions otherwise out of reach. If you can do this by yourselves, so much the better. The same goes for the valuable services of an electronics technician. This is not to say that one should not at least be conversant with some of the more widespread electronic devices and their circuits, since the detection of and manipulation of signals in the measuring process will invariably demand this. One might be led into believing that the advent of data acquisition system sensors (usually referred to as MBL, an acronym for microcomputer-based laboratory) has obviated the necessity for such skills, and to a certain extent this is so. Nevertheless, employing only these will restrict the scope of measurements to their regime of operating characteristics, without a hope of expanding one's horizons in order to minimize noise, to match sensors and transducers, to shape and amplify signals (every measurement can be converted to a voltage), reduce distortions, perform a series of functions in tandem, and so on. The electronic instruments now available can perform many functions, but it is fair to say that few people know how they work in any detail; this need not be a problem, so long as they do the job and can be calibrated. There is some advantage to using simple analog instruments (analog oscilloscopes, signal generators, multimeters, etc.), particularly for people who are just beginning. There may be mystery in the project, but why have it also in the instruments?

Hoarding the good stuff

Budgets are usually tight, but there are some basics without which one is not in business. Power supplies, signal generators, oscilloscopes, optical and electronic components are among these. More of this will be discussed in Appendix B on instrumentation. But appetite comes with the eating and, as you go along, you learn that serendipity favors those who are aware of needs. So keep an eye on equipment discarded by researchers in academic and industrial institutions, and keep in touch with them – for projects in schools one does not need, or even want, the last word in sophistication, and the tools of the previous generation, or the last but one, are just fine. Bureaucracy often dictates that equipment not be given to you but only lent for a limited but often indefinite period. Who cares?

Know where electric motors are reconditioned, for there you will find motors of all sizes, cheap or free, and insulated wires of all diameters for making coils. Use DC motors as speed sensors. Look out for amplifiers from audio equipment, microphones, loudspeakers from defunct radio and television receivers, magnets around the magnetron in microwave ovens, step motors and fans from junked computers – the list is long. Never buy anything new without searching the web and eBay; the supply and choice there are immense.

Projects are about and with materials, so you need to learn where your area's stores are situated for lumber, building materials, metal, plastic, Plexiglas (Perspex), glass, pipes, hardware (ball bearings, nuts and bolts, fasteners and glues, electrical parts), and electronics parts. The more project-crazy you become, the more expert you will be in locating what you need, for needs multiply as the horizon is extended. Be aware, and let your eyes roam in these stores, to know what gadgets and materials are around and might come in handy in the future. More details are given in Appendix B.

Reference material

Our researchers and we guides always have much to learn as we go along. Often a guide is not an expert in the particular field of physics needed for the project. Looking up or learning theory, physical constants, material parameters, and mechanical and electromagnetic data will be a constant need. Much can be found on the web, but in our experience there is no substitute for first-class texts and monographs when one wants to learn about a topic in depth. A university or college library will usually have the basic needs, but can one suggest a useful school library? At the end of this book we list some of the accessible books that we have found useful. We suggest that sooner or later these should find their way to the shelves of all readers of this book, together with handbooks and catalogs.

Design, invention, and perseverance

We have little to say on these necessities. The hope is that some of the researchers will be more inventive than the guide (we all wish that our children and researchers turn out better than ourselves). We realize that necessity is the mother of invention, but what if the latter is delayed beyond endurance? Discuss things with your colleagues, they may have an idea. Then there are the experts in their fields at the nearest college or university. Coming to them prepared with a well-defined problem and situation, and an account of your failed attempt to solve it, can often elicit a sympathetic response and result in useful advice.

Do not abandon an investigation at the first serious obstacle – this results in a feeling of letdown for all. We know, because we have done this and tasted the bitter pill of failure. Modify, branch out in another direction, fight, and persist. Something of value will turn up.

Researchers, choices, and the toil

We have emphasized that projects are for the benefit of both guides *and* researchers. But the situation is not symmetric, the burden of responsibility for steering the project falls on the instructor.

Partners

The best hope is that researchers will choose partners among themselves, but the instructor may be called on to sort out difficult or pathological cases. As always in these matters, listen, be understanding, and try to help. The pairing of partners with parallel interests by means of a simple questionnaire often works.

Choosing a topic

If researchers have a topic that fascinates them, so much the better. Popular subjects are sports, music, flight, human perception. Some students come with a firm choice in mind – others seem not to have any idea. Sometimes giving examples of past projects classified under mechanics, heat, sound, electromagnetism, light, atoms may inspire choice. But some of these may turn out to be so general, outlandish, or expensive that some focusing and paring down may be necessary, some negotiated settlement achieved which still leaves the embers of their enthusiasm glowing, and the feeling that the idea was theirs. Preservation of such motivation will contribute to steadfastness and a better chance of success. Here are two examples of negotiated reductions of aspirations.

A researcher wanted to work on telemetry and rocket propulsion. He finished up building a water rocket, rigged up a force probe to an analog to digital (A/D) interface to measure reaction at lift off, and devised an optical method for measuring the maximum height reached.

Another researcher was interested in interior lighting. This led to an investigation of the dependence of incandescent bulb lifetime on voltage, with excursions into radiation and thermodynamics.

Not every researcher knows what to choose, nor are all their choices viable. In any case, a deadline must be announced by which vacillations come to an end and choices, or the obvious lack of ideas, are clear. That also means the guide must have a reservoir of ready-made suggestions, pet ideas that have been waiting for a suitable pair to explore. Past projects that have only come to partial fruition and have much potential for further development may well be among these.

The daily toil

Carrying out a project requires many skills and aptitudes, for researcher and guide alike: planning, manual dexterity, building, mathematical analysis, verbal articulation, wonder, enthusiasm. One or more may be present to an insufficient degree in a particular person. For this reason too, working in pairs is necessary, whereby the partners complement each other. In the periodic meetings with the researchers the guide will soon diagnose deficiencies in one or the other's skills and suggest ways to improve them. Among our drives, that towards competence is a major one, and one of the yardsticks for success as the projects of the whole class proceed, and a source of satisfaction for the guide, is a perceptible improvement in some area of competence for every researcher.

Guiding projects differs in many ways from giving a lecture course, and is more like tutoring graduate researchers. The following features characterized our programs, sharply differentiating them from the usual mode of instruction: any problem mutually agreed upon by guide and researcher could be investigated, no clear-cut final answers were known in advance, work was done in blocks of uninterrupted time, there was plenty of planning but no syllabus, there were exercises to learn how to use equipment, there was time to think and discuss matters outside the laboratory.

At the Massachusetts Institute of Technology (MIT), projects lasted a semester, working in scheduled weekly afternoon sessions of five hours. The Project Lab course was offered in each of two semesters a year, and it ran for 30 years as a scheduled course. Some 22 ± 2 students participated in each semester, accompanied by a guide and an assistant.

At the Israel Academy of Sciences, a high school for the gifted, Project Lab has been conducted for 20 years and is continuing. The weekly allotment was a two-lesson period a week, but lasted throughout the year, with the added advantage of a boarding school, which means that researchers could come and work outside scheduled hours, the lab would be open in the evening once or twice a week, and a lab assistant would be present. Two guides were assigned to some 20 researchers, each responsible for 5–7 projects, each benefitting from the support of an experienced and enthusiastic lab assistant.

The time burden may be heavy, both for guides and researchers. If possible, the guide also needs to be available for consultation outside the time slot officially allotted to the projects. For continuity it is vital to have a laboratory assistant who accompanies the projects year after year, is familiar with the equipment and materials available, and shares some of the qualities and skills that are required of the guides, namely, ingenuity, experience, broadness of knowledge, and interests. These may not always be present even with the best of teachers or researchers.

Some kind of closure for the whole class is necessary. This includes a careful write-up of the project. In addition, it is useful to have a private hearing of each team, devoted to a summary and questioning by the guide and a colleague, maybe even in the presence of an external examiner, lasting say 20–30 minutes.

This may also serve as a preparation for a short lecture to be given, the audience being the rest of the class and possibly some invited guests.

Is it worth the effort?

The benefits to the guide are many. There is no better way to get to know researchers than in guiding them in projects. For guides at high schools or colleges where research time is absent or minimal, projects are an ideal way to get refreshed, extend our own knowledge and skills, and break the routine of lecturing to a class. We might well adopt some of the materials and findings of a project in our daily teaching, thereby making it more alive and attractive. For guides active in research, guiding projects is an opportunity to share their skills and knowledge, a laudable and necessary community service, with the side benefit of possible recruiting of future undergraduate or graduate research students.

The researcher will gain independence and learn a little about how science is done. She or he will achieve something that was the fruit of struggle and labor, and take some pride in it. It may also be an influence in considering science as a career, something we science teachers are surely interested in promoting.

About the book

We describe projects to suit various tastes and resources in high schools and colleges. Separate parts are devoted to mechanics, electricity, acoustics, liquids, optics, and heat. Appendix A contains a list of additional ideas for projects, without the details given in the rest of the chapters. Some of the suggested projects may contain several of the above-mentioned topics. Theoretical background and summaries are given in those cases where we felt that they would not be readily available, but we frequently refer readers to the original articles in the teaching journal literature or books where more details can be found. Discussion of the original articles between advisor and researcher is a vital part of the process.

We hope that by suggesting goals and possibilities for each exploration we have set the stage and expectations for the reader. Not all those listed need or can be realized within the time or with the equipment available. It may well be that guide and researcher will suggest their own modifications and extensions, in which case we hope that they will inform the learning community at large, for the benefit of all.

There will often be alternatives to the methods, detectors, instrumentation, and data processing suggested in the section on experimental suggestions. Budget considerations will no doubt play a role in defining standards of sophistication and accuracy, but we have usually set these to a level commensurate with the equipment of undergraduate or high school laboratories. Some guides might well think that there is much value in researchers building their own sensors, A/D converters, control circuits, or software for data analysis, depending on the expertise of those involved and the skills the guide wishes to nurture and develop for his charges. This is indeed a laudable aim and part of the philosophy of Project Lab: minimize mystery. But the universal presence of complex, sophisticated digital devices in the lives of researchers means that there should be no barrier to using similar devices in projects. The ready availability of packages for data acquisition and analysis, digital oscilloscopes with memory, fast Fourier transform programs (FFT), and spectrum analyzers has made the carrying out of fairly extensive experimental investigations available to a wider audience, and many of the projects in the book rely on such facilities.

What follows are descriptions of projects that have been invented and done, and are examples for the reader-guide to build on with his or her researchers. Appendix B should help in providing materials and setting up facilities, some necessary at the start, some that evolve with each session.

REFERENCES

1. G. Lockett, in *LabNet: Toward a Community of Practice*, R. Ruopp (ed.) (Erlbaum Associates, Hillsdale, NJ, 1993), p. 351.
2. P.M.S. Blackett, The craft of experimental physics, in *University Studies*, H. Wright (ed.) (Nicholson & Watson, London, 1933).
3. F. O'Brien, private communication.

Part 1
Mechanics

1	Bouncing balls	13
2	Mechanics of soft springs	17
3	Pulse speed in falling dominoes	25
4	A variable mass oscillator	28
5	Rotating vertical chain	34
6	Cycloidal paths	38
7	Physics of rubber bands and cords	44
8	Oscillation modes of a rod	49

Part Two

Mechanics

Bouncing balls

Introduction

Bouncing plastic or rubber balls inflated to different pressures present the experimenter with an array of parameters, and it is a challenge to be able to isolate one or the other in a controlled manner in order to measure its influence on what happens in a bounce. Interesting effects can also be investigated when non-inflatable balls (ball bearing, baseball, golf, super, or tennis balls) are dropped on various surfaces, such as granite, steel, or wood. Among these are coefficients of restitution, static and dynamic hysteresis curves related to energy losses, and the time and distance dependence of the ball–surface interaction.

Introduction	13
Theoretical ideas	13
Goals and possibilities	14
Experimental suggestions	15

Theoretical ideas

Inflatable balls

After first contact with the surface the ball continues to move down a certain distortion distance d, before coming to a stop and rebounding. Energy and momentum considerations enable one to estimate, for a given drop height h, both d and the average force F of the surface on the ball, the time τ for which the ball remains in contact with the surface, and the deceleration after contact. Since both the area of contact S for maximum compression and τ can be measured, it is natural to ask how τ depends on impact speed, and how S and τ depend on h, on the excess pressure P in the ball, and on the radius of the ball R.

A very crude model for the distortion can give some guidelines and a point of departure for comparison with measurements [1]. If one neglects the stiffness of the ball's material, the contact force is S times P in equilibrium. The simplest assumption for S is to assume that the rest of the ball retains its spherical shape but that the bottom forms a circle of radius a, as in Fig. 1.1. Then $d = R - b$, $a^2 \approx 2Rd$ (for d small, $d \ll 2R$), and so $F = (2\pi RP)d$: The force is proportional to the displacement, the "force constant" being $2\pi RP$. Equating the elastic potential energy to the initial kinetic energy on impact one finds that

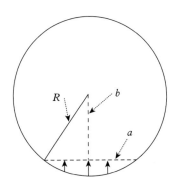

Figure 1.1 *Ball model.*

$$d = (mgh/\pi RP)^{1/2}, \quad (1.1)$$

where m is the mass of the ball. The contact area S is

$$S = \pi a^2 = 2\pi Rd = v(2\pi Rm/P)^{1/2} = (4\pi Rmgh/P)^{1/2}, \quad (1.2)$$

where v is the speed of impact. In this oscillator model the contact time τ is at least half the oscillation period, so that, approximately,

$$\tau = (\pi m/2RP)^{1/2}. \tag{1.3}$$

The contact time in this model does not depend on the impact speed v.

Of course, reality may well be very different: the more the ball is compressed, the more it may resist further deformation, and a single force constant is inadequate to describe the dynamics. The contact force will therefore not be proportional to the deformation distance d. A number of further effects have been neglected in the model. First, on impact, surface waves are excited in the wall of the ball, and carry away energy, so that equating the elastic with the kinetic energy certainly needs to be reconsidered. The surface wave energy should be smaller for higher inflation pressure, when the distortion is smaller. Constructing a theory to take these waves into account would by itself constitute a complicated project. Secondly, the pressure on distortion will not be the equilibrium pressure P, because of the volume change. This should be important for large v and low inflation pressure. Finally, there is the frictional loss between ball and surface.

In order to gain understanding of the real deformation of the ball on impact, a computer dynamical simulation has been performed by dividing the ball into a large number of segments, calculating the pressure and surface tension forces on each, and using these to calculate the shape of the ball as a function of time. With a couple of adjustable parameters the simulation obtains excellent agreement with experiment for the contact time and the contact area as a function of impact speed. The interested reader is referred to the original paper for details [2].

Coefficient of restitution

There are various models for the velocity dependence of the force amplitude F_0 and of the duration of contact τ. In the Hertzian model for colliding spheres [2], $F_0 \propto (v)^{1.2}$ and $\tau \propto (v)^{-0.2}$. These can be checked experimentally.

One can monitor several jumps of the balls on a particular surface. At each jump the coefficient of restitution e is v_f/v_i, where v_f, v_i are the velocities just after and before the collision. After the nth jump, $e_n = v_{n+1}/v_n = t_{n+1}/t_n$, where t_n is the time of flight of the ball between the nth and $(n+1)$th jumps. A plot of e_n versus t_n would reveal the constancy or otherwise of e_n.

Goals and possibilities

- Measure the contact area as a function of impact speed for various inflated pressures.
- Measure the contact area as a function of inflation pressure for a particular impact speed.
- Measure the contact time for various impact speeds and pressures.

- Measure the ratio of rebound height to the initial drop height as a function of impact speed.
- Measure the static force as function of distortion distance d for a particular pressure, and at the same time measure the pressure change inside the ball as the ball is squashed.
- Measure the force on the ball at impact as a function of time.
- Investigate the time dependence of the force F exerted by various surfaces on the various balls.
- Figure out the dependence of F on the distance y descended by the ball after the moment of contact. Plot the force versus distance curves and from these deduce energy losses and force laws during collisions.
- Compare these dynamic curves with static curves obtainable by compressing the balls between plates over a period of one minute, holding, and then releasing the stress over one minute.

Experimental suggestions

A 20 cm diameter hollow plastic inflatable ball found in most toy stores is convenient to work with.

Rebound height can be measured by gradually raising on a jack a laser whose horizontal beam will be just intercepted by the ball, allowing accuracy to within a millimeter.

Contact area can be measured by letting the ball fall on a thin aluminum foil covering sandpaper, and measuring the area of the imprint made by the ball on it upon impact.

Contact time may be measured by building a device whereby the ball, made conducting by covering it with conducting foil, mylar, or conducting paint, closes a circuit on the floor as it falls on it. As shown in Fig. 1.2, this activates a scaler-timer as long as there is contact.

Contact area and contact force can be measured as a function of distortion distance by controlled static squashing of the ball in some device. One possible solution for doing this is shown in Fig. 1.3. Using a force platform transducer connected to an MBL and sampling at a suitably high rate would give interesting insights into the contact force $F(t)$ as a function of time as the ball lands on it. The integral of this force is the impulse, equal to the change in momentum which can be measured. Look for possible asymmetry in the force versus time curve and interpret it.

Solve numerically the equation of motion for the ball's center of mass y from time zero at first contact with the surface, $y'' = F/m$, subject to the initial conditions: $y = 0$, $y' = v_i$ at $t = 0$. This enables one to plot F as a function of y for the total collision time and discover whether and how much hysteresis there is in the collision (the loop enclosed by the F versus y curve), which should be equal

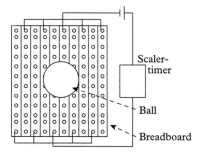

Figure 1.2 *Measuring contact time.*

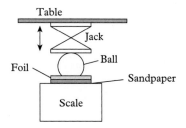

Figure 1.3 *Measuring contact area and contact force.*

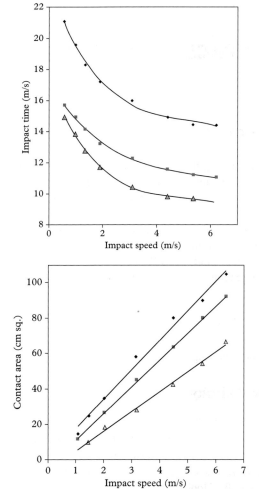

Figure 1.4 *Contact time as function of impact speed: diamonds 17 kPa, squares 36 kPa, triangles 98 kPa.*

Figure 1.5 *Contact area as a function of impact speed: diamonds 17 kPa, squares 36 kPa, triangles 98 kPa.*

to the energy lost: $0.5m(v_i^2 - v_f^2)$. It is also of interest to see whether the curve has any portion that is a straight line, where Hooke's law would hold, or whether some other (Hertzian [3]) law holds.

Some typical results of experiments, reproduced from Ref. [1], are shown in Figs 1.4 and 1.5, which show contact time and contact area as a function of impact speed, for various pressures.

....

REFERENCES

1. N.J. Bridge, The way balls bounce, *Phys. Ed.* **33**, 174 (1998).
2. N.J. Bridge, The way balls really bounce, *Phys. Ed.* **33**, 236 (1998).
3. G. Barnes, Study of collisions, *Am. J. Phys.* **26**, 5 (1958).

Mechanics of soft springs

Introduction

Very soft springs, among them the slinky toy, have unusual properties which are different from springs normally used in teaching laboratories: when suspended under their own weight their extensions are enormous compared to the unstretched length; they need not have any pre-stressed tensions in them; for small loads they do not obey Hooke's law; the periods of vibration, whether or not the spring is loaded, do not obey the usual formula; pulse velocities along them are dependent on position; and so on. These properties make these springs interesting objects for investigation.

Introduction	17
Theoretical ideas	17
Goals and possibilities	20
Experimental suggestions	20

Theoretical ideas

For details on the following ideas, see Refs [1–8].

The vertical slinky

Longitudinal pulse velocity

See Ref. [5]. Let L be the length of a freely suspended slinky, and $h(x)$ the distance between successive turns at position x, as in Fig. 2.1, which also shows the turn number n. Let κ and m be the force constant and mass of a *single* turn, respectively. By Hooke's law for a single turn, the tension $S(x)$ at x is $S(x) = \kappa h(x)$, assuming there was zero separation between the turns in the undistorted state. The mass per unit length is $\mu = m/h(x)$, so that from the usual formula, $v = \sqrt{(S/\mu)}$, the speed of a longitudinal pulse is

$$v(x) = h(x)\sqrt{(\kappa/m)}, \tag{2.1}$$

and is thus position dependent.

To find $h(x)$, note that the tension $\kappa h(x)$ is the result of supporting the weight of all the turns below x. Since the number of turns in the length dx is $dx/h(x)$, the total number of turns from x to L at the bottom is $\int_x^L dy/h(y)$, and each has a mass m. Therefore, we have

$$\kappa h(x) = mg \int_x^L dy/h(y). \tag{2.2}$$

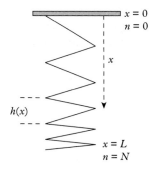

Figure 2.1 *Vertical slinky.*

This may be differentiated to give $\kappa\, dh/dx = -mg/h(x)$, whence integrating will give

$$h(x) = [2mg(L-x)/\kappa]^{1/2}. \tag{2.3}$$

The velocity is given by combining (2.1) and (2.3):

$$v(x) = \sqrt{2g}\sqrt{(L-x)}. \tag{2.4}$$

An indirect way of checking (2.4) would be to measure the total time τ for a pulse to travel from the lower end, to be reflected at the top and arrive back to its starting point. From (2.4) we have

$$\tau = 2\int_0^L dy/v(y) = 4\sqrt{L/2g}. \tag{2.5}$$

But since from (2.1) the time h/v to traverse one turn is a constant equal to $\sqrt{m/\kappa}$, then for the round trip of $2N$ turns the time τ may also be written as

$$\tau = 2N\sqrt{m/k}. \tag{2.5a}$$

If a force F acts downward at the bottom of the slinky, the following modifications to the above apply: F is added to the right-hand side of (2.2), (2.3) becomes

$$h(x) = [(F/\kappa)^2 + 2mg(L'-x)/\kappa]^{1/2},$$

and $L \to L'$ (the stretched length) in the upper limit of the integral in (2.5), but the time τ in (2.5a) is independent of F.

Equilibrium configuration

This may be derived from (2.3). In the absence of F, at the top ($x = 0$) we must have in equilibrium that the tension $\kappa h(0)$ equals the weight of the spring $Mg = Nmg$, where N is the total number of turns. Substituting this into (2.3) and rearranging, we find a parabolic relationship between the length and the number of turns:

$$L = (mg/2\kappa)N^2. \tag{2.6}$$

A graph of L versus N^2 should be a straight line of slope $mg/2\kappa$, thus enabling us to determine κ.

Standing waves

In the presence of waves, the equilibrium turn-spacing $h(x)$ will be extended by an additional amount $u(x,t)$. The wave equation is derived from Newton's second law for a single turn: the net force on a single turn must be equated to $m\partial^2 u/\partial t^2$. But the net force is just the *difference* of the extra force $\kappa(\partial u/\partial x)h(x)$ caused by $u(x,t)$ on either side of the turn, namely, $(\partial/\partial x)[\kappa h(x)(\partial u/\partial x)]h(x)$, so that, dividing by $h(x)$, we obtain

$$(\partial/\partial x)[\kappa h(x)(\partial u/\partial x)] = [m/h(x)]\partial^2 u/\partial t^2, \tag{2.7}$$

where m/h is the linear mass density.

If we use the (discrete) turn number n instead of x as our variable, then because $dx = h(x)dn$, we may write $\partial/\partial x = (1/h)\partial/\partial n$, which immediately simplifies (2.7) to the usual wave equation

$$\frac{\partial^2 u}{\partial n^2} = \frac{m}{\kappa}\frac{\partial^2 u}{\partial t^2}, \tag{2.8}$$

in which $\sqrt{\kappa/m}$ (= const.) is the speed in *turns* per second. This is an interesting result: the wave covers the same number of *turns* per second, though not the same *distance*, since $v(x)$ is linear in x.

The rest of the analysis proceeds as usual, via separation of variables for the wave equation and standing waves, together with boundary conditions which will determine the normal mode frequencies. These are $u(n = 0, t) = 0$ and $\partial u/\partial n = 0$ at $n = N$ for a free bottom end, resulting in the possible frequencies

$$f_{r,\text{free}} = \frac{2r + 1}{4N\sqrt{m/\kappa}} = (2r + 1)/2\tau. \tag{2.9}$$

If also the bottom end is fixed then u vanishes there and so

$$f_{r,\text{fixed}} = \frac{r + 1}{2N\sqrt{m/\kappa}} = (r + 1)/\tau, \tag{2.10}$$

with $r = 0, 1, 2, \ldots$. Thus the fundamental frequency for the fixed end is twice that for the free end.

Vertical soft spring

For small loads m_1 hanging from a vertically suspended soft regular spring, satisfying $m_1 g < S_0$, where S_0 is the built-in compression (negligible for a slinky, small for a very soft regular spring), the tension–extension relationship is parabolic, whereas for $m_1 g > S_0$ it is linear, as shown in Ref. [2]. The two regimes join smoothly.

Horizontally suspended slinky

It is of interest to measure the velocity of pulses, both transverse and longitudinal, along the horizontal slinky. This velocity is a constant [7,8]

$$v = \sqrt{(kL/\mu)}, \qquad (2.11)$$

where μ, k, L are the mass per unit length, the force constant, and the stretched length of the slinky, respectively. Since $\mu = M/L$, it follows from (2.11) that the time for a round trip,

$$\tau = v/2L, \qquad (2.12)$$

is independent of L.

Goals and possibilities

- Measure the length as function of the number of turns for a vertical slinky.
- Obtain or make a soft spring which has a low enough built-in compression, so that the stress–strain (or length–weight) curve can be explored for m_1 very small, the object being the exploration of the possibility that the behavior will not be according to Hooke's law.
- Measure the period of oscillations of the spring as function of loading, including oscillations under its own weight.
- Generate longitudinal vibrations in the spring stretched vertically to various lengths between a mechanical vibrator and a post, and measure the period of vibrations for the fundamental and higher modes.
- Large slinkies can stretch to 10–15 m when hung vertically. A pulse sent from the top clamped end will slow down as it proceeds downward. Devise a way to measure this variation of pulse speed.
- Measure force constant, pulse velocity, and round trip time for a horizontally suspended slinky.

Experimental suggestions

Soft spring tension versus extension

One needs a very soft spring, a set of small weights (these may be made from circular washers with a gap so that they could be hung on a home-made holder; see Fig. 2.4), and some device to measure extension: a sonic ranger placed on the floor directly below the weight holder can measure the diminishing distance from it as the spring is loaded.

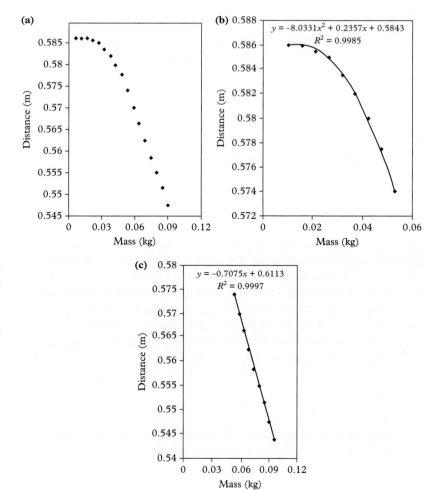

Figure 2.2 *Linear and parabolic behavior of soft spring.*

The results [9] in Fig. 2.2(a) show an initial parabolic section, followed by linear behavior. Drawing separate plots for the two regions in Figs 2.2(b) and (c) makes this clear. The first section is fitted well with a second degree polynomial and the second with linear regression.

Equilibrium configuration of a slinky

One can suspend the slinky from any turn along its length, let the rest dangle freely, and measure the length $L(n)$ of the n free turns. Figure 2.3 shows $L(n)$ for a small slinky of 90 turns, diameter 5 cm, and undistorted length 5.5 cm, stretching to 60 cm when hanging freely. The quantity κ may be calculated from the equation of the trendline.

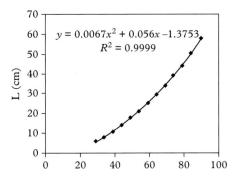

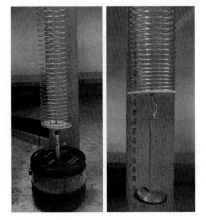

Figure 2.3 *Length L as a function of winding number n.*

Figure 2.4 *Attachments to slinky.*

Normal modes of a slinky

Choose the small slinky from the previous experiment. Figure 2.4 shows a way to solve the problem of attaching to the bottom centrally either weights or the vibrator, and of monitoring the oscillations: glue a 1 mm thick disk made of balsa wood to the slinky bottom. The center of the disk has a hook driven through, to which weights or a vibrator can be attached. The weight holder has a platform on which 5 gram weights may be hung. A sonic ranger whose pulses are reflected from the bottom of the platform records the oscillations, shown in Fig. 2.5, their trace supplying the period and the frequency. Care must to be taken to record only the steady state and not transient oscillations. In addition, it is crucial to prevent coupling to sideways motion as much as possible.

Unlike for the case of a vibrating string, the *spatial* distribution of nodes is not uniform, but it *is* uniform when expressed in terms of the turn number n, counted from the point of attachment ($n = 1$ at the top, $n = N$ at the bottom). Thus, for the rth mode the number of turns Δn between each mode is $N/(r+1)$. This may be

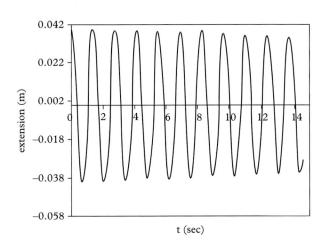

Figure 2.5 *Sonic ranger output for period.*

confirmed by photographing the standing waves and simply counting the turns between the nodes. Do this by dabbing all the turns with a yellow highlighting marker containing fluorescent paint. An ultraviolet source shone on the slinky and a digital camera recording in the dark, with a suitable grid behind the slinky to provide the scale, will then do the trick.

Velocity of pulse as a function of position

See Ref. [9]. Choose a very large slinky (say with 190 turns, diameter 8 cm and unstretched length 12.5 cm, stretching to 13.4 m when hung vertically). Hang it from the highest roof of your building, say 15 m above ground level. Generate a pulse and photograph its journey from one end to the other and back with a digital camera recording at 60 frames a second. Background contrast has to be provided (hang a black drape behind the slinky), equipped with horizontal light strips of masking tape stuck across it at regular intervals. This provides one with the vertical coordinates needed when analyzing frame by frame the progress of the pulse along the slinky. Figure 2.6 is a plot of the velocity as a function of $\sqrt{(L-x)}$. Because of the coarse grid used, in the analysis of the video frames only average values of the velocity of the pulse can be calculated. Agreement with the theory is reasonable: a 3% deviation for the slope of the best straight line from the theoretical value of $\sqrt{(19.6)}$. This experiment will certainly attract the attention of many onlookers from the school.

Horizontal slinky

Suspend the slinky with long threads attached to every third turn, in order to minimize the hindrance to the passage of the pulse and the appearance of spurious transverse forces. The other ends of the strings are attached to smooth plastic rings which can be slid easily along a horizontal bar, as shown in Fig. 2.7. One end of the slinky is fastened to a stand, the other attached to a pulley over which the string passes, at the end of which different weights may be hung when measuring the force constant or the pulse velocity for different stretches. It is vital to arrange

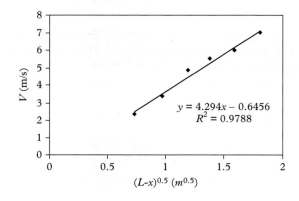

Figure 2.6 *Position-dependent pulse speed.*

Figure 2.7 *Horizontal supports for slinky.*

the strings to be perfectly vertical and to be equidistant from each other all along the length of the slinky. Follow the progress of a pulse by a video recording and analyze the frames to prove (2.11) and (2.12).

....

REFERENCES

1. A.P. French, The suspended slinky: A problem in static equilibrium, *Phys. Teach.* **32**, 244 (1994).
2. G. Lancaster, Measurements of some properties of non-Hookean springs, *Phys. Educ.* **18**, 217 (1983).
3. T.C. Heard and N.D. Newby, Behaviour of a soft spring, *Am. J. Phys.* **45**, 1102 (1977).
4. A.P. French, *Vibrations and Waves* (W.W. Norton, New York, 1971), p. 61.
5. R.A. Young, Longitudinal standing waves on a vertically suspended slinky, *Am. J. Phys.* **61**, 353 (1993).
6. T.W. Edwards and R.A. Hultsch, Mass distribution and frequencies of a vertical spring, *Am. J. Phys.* **40**, 445 (1972).
7. E. Huggins, Speed of wave pulses in Hooke's law media, *Phys. Teach.* **46**, 142 (2008).
8. G. Vandegrift, Wave cutoff on a suspended slinky, *Am. J. Phys.* **57**, 949 (1989).
9. P. Gluck, A project on soft springs and the slinky, *Phys. Educ.* **45**, 178 (2010).

Pulse speed in falling dominoes

3

Introduction	25
Theoretical ideas	25
Experimental suggestions	26
Further explorations	27

Introduction

A row of falling dominoes, initially upright, may be used to investigate the speed of a pulse along it, when the first one is given a push with low initial angular velocity. For a small number of dominoes the speed may depend on the number. For large numbers, it is the initial spacing of the dominoes and their height which influence the speed of the pulse. Photogates can be used to measure times accurately. The project provides opportunities for employing various model calculations to fit the experimental measurements, ranging from simple scaling to numerical solutions of the equations of motion in rigid body dynamics.

Theoretical ideas

Scaling

Let d, h, t, and w be the spacing, height, thickness, and width of the dominoes, with $t \ll h$, as shown in Fig. 3.1.

We may reasonably assume that the domino width w does not appreciably affect the motion and that, except for small separation, the domino thickness t plays no role.

The wave speed V depends on d, h, and g (acceleration due gravity), so $V = V(d,h,g)$. From these three quantities one can form the dimensionless ratio

$$V/\sqrt{gh} = f(d/h). \tag{3.1}$$

Figure 3.1 *Geometry of stacked dominoes.*

Thus the normalized speed depends on the relative spacing. This relationship may be tested experimentally by plotting $V/\sqrt{(gh)}$ versus d/h for two different heights, in the full range of values for d/h: both when d/h is comparable to t/h, in which case the row falls almost like a solid body, and when d/h is close to one, when the dominoes will not fall over straight, so that scatter of data is expected.

Figure 3.2 shows the results for 100 dominoes with $h = 4.45$ cm, $t = 0.79$ cm, $w = 2.23$ cm. Two heights were tested, h and $2h$, the double height (the single square point in the graph) being obtained by taping two dominoes together. For relative spacing approaching the domino thickness-to-height ratio,

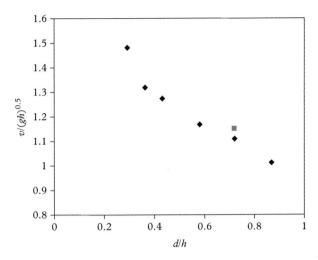

Figure 3.2 *Testing equation (3.1).*

$d/h \to t/h$ (here $t/h = 0.17$), the normalized wave speed, the uppermost experimental point in the graph, is near the solid body value $V/\sqrt{(gh)} = \sqrt{2}$ (i.e., the value you would get for any falling object by equating the decrease in potential energy to the increase in kinetic energy). Various fits may be tried to the experimental data points.

A model calculation

A computer simulation [1] of the problem can be performed for the fall of 20 dominoes. In treating the motion of a single domino from its vertical position to its impact on the next one in the row, it takes into account the effects of the motion of all other dominoes. It uses: (a) energy conservation between collisions to calculate instantaneous velocities, (b) conservation of angular momentum for a collision to calculate a new set of angular velocities after contact, and (c) geometry to relate the angular positions and angular velocities at any instant of time. The time interval between the first domino striking the second one and the nth one striking the $(n+1)$st one is calculated from the angular velocity of the nth one by integration. We refer the reader to the original article [1] for details, which also suggests a refined model that would take into account energy dissipation, as the dominoes slide over each other. Such computer simulations would be a valuable addition to the experimental investigation.

Experimental suggestions

Two different investigations are of interest:

1. A small number N of dominoes (or uniform wooden bricks from children's building block sets) standing upright on their narrow edges are placed as

a linear chain at fixed uniform spacing d, on a rough surface of, say, fine sandpaper. The latter ensures that the point of contact between the falling domino and the surface is at rest. Attach a short piece of blackened opaque thin drinking straw to the first and last dominoes in the row. These are positioned so as to interrupt appropriately placed photogate beams on falling. The first domino is released from a position of unstable equilibrium, and the time $T(N)$ between the first and last domino's fall is measured. As N increases, say between 4 and 20, $T(N)$ increases with it. The results may then be compared with those of computer simulations [1].

2. For a fixed large number of dominoes (between 100 and 150), the pulse speed is measured as a function of the separation d between the dominoes. The height of the dominoes h may be doubled by sticking one on top of another. It is interesting to compare speeds for the same number of dominoes and for the same spacing, but for two different heights. Once again, the first and last dominoes in the row are provided with a drinking straw, interrupting respective photogates for timing. Times can be measured with a data acquisition system, or with a scaler-timer. Typical scaled data for two different heights are shown in Fig. 3.2.

Further explorations

An extension, or a separate project, could be devoted to theoretically predicting and experimentally measuring the time it takes for a chain of dominoes to fall, as a function of N, d, the total distance spanned by the chain, and the shape of the chain. These questions are addressed in Refs [2–4], which also provide extensive bibliographies.

....

REFERENCES

1. D.E. Shaw, Simulations for falling dominoes, *Am. J. Phys.* **46**, 640 (1978).
2. C.W Bert, Falling dominoes, *SIAM Review* **28**, 219 (1986).
3. G. Catlin, S. Fields, and L. Mullen, Falling dominoes, *Am. J. Phys.* **57**, 1089 (2005).
4. J.M.J. van Leeuwen, The domino effect, *Am. J. Phys.* **78**, 721 (2010), also online: <http://www.stat.physik.uni-potsdam.de/~pikovsky/teaching/stud_seminar/domino.pdf>.

4 A variable mass oscillator

Introduction 28
Theoretical ideas 28
Goals and possibilities 31
Experimental suggestions 31
Further explorations 33

Introduction

This project investigates an oscillator whose mass decreases linearly with time. It provides an opportunity to evaluate the relative importance of two velocity-dependent models for air resistance influencing the motion. Since the mass decrease is due to the efflux of sand from a container, the work has some relevance to the physics of granular materials. The latter topic could be a source for a number of projects (see Appendix A).

Theoretical ideas

The oscillating body is an inverted narrow-neck bottle suspended from a spring which itself hangs from a force sensor. Sand issues from the opening, whose size can be varied. The rate of flow is independent of the amount of sand remaining – a property characteristic of granular materials [1], but unlike a regular fluid. Let $m(t)$ be the mass of the suspended body at time t. Since the mass loss rate is a constant, say c, one may write

$$m(t) = m_0 - ct, \qquad (4.1)$$

where m_0 is the initial mass. To make the analysis tractable, we shall assume that, relative to the bottle, the exit velocity of the sand at the orifice is negligible. As the bottle oscillates on the spring, it experiences air resistance whose value $R(v)$ depends on the velocity v of the body. There are models in which this dependence is linear, or quadratic, or both. Let us assume then that

$$R = -bv - av^2, \qquad (4.2)$$

where a and b are constants. The coordinate of the bottle is x, positive down, and $x = x_0$ at the equilibrium position, where $kx_0 = m(t)g$, k being the spring constant. The equation of motion of the system will be

$$m(t)\ddot{x} + a|\dot{x}|\dot{x} + b\dot{x} + kx = 0. \qquad (4.3)$$

The experiment will be guided by the analysis of this equation. We shall present here only an outline and the final results, and refer to the literature for more details [2,3]. Of interest is the time dependence of the amplitude of oscillations, $A(t)$, as well as $x(t)$. We consider two models for $R(v)$.

a = 0

We can convert (4.2) into an equation for the energy variation (loss) in time due to air resistance and mass outflow by multiplying by dx/dt, writing $E_p = \frac{1}{2} kx^2$ for the elastic potential energy, $E_k = \frac{1}{2}mv^2$ for the kinetic energy, and using $m(t)$ in (4.1):

$$d(E_k + E_p)/dt = -(b + c/2)(dx/dt)^2. \tag{4.4}$$

We may use (4.4) in order to follow the dependence of the amplitude on time, $A(t)$. If we consider two consecutive oscillations we may write for the rate of energy loss:

$$d(E_k + E_p)/dt = d/dt(kA^2/2) \approx -(b + c/2)\text{av}(v^2), \tag{4.5}$$

where we used the mean square velocity $\text{av}(v^2)$ in any period rather than instantaneous values, approximately equal to $A^2\omega^2/2$ (ωA is the maximum velocity), and the period T is given by the usual formula $T = 2\pi/\omega$. We shall also borrow the usual expression for a constant mass oscillating on a vertical spring, namely $\omega^2 \approx k/m(t)$. This relation is borne out by experiment. With these approximations we may write (4.5) as follows:

$$dA/dt \approx \frac{1}{2}\left(b + \frac{c}{2}\right)\frac{A}{m(t)}. \tag{4.6}$$

The solution for $A(t)$ becomes:

$$A(t) = A_0\left(1 - \frac{ct}{m_0}\right)^\Delta, \tag{4.7}$$

$$\text{where } \Delta = b/2c + 1/4. \tag{4.8}$$

It is interesting to consider three special cases:
(i) $c \to 0$: In this case,

$$A(t) = A_0 \exp\left(-\frac{b}{2m_0}t\right). \tag{4.9}$$

This result resembles that for the decay of an underdamped oscillator. Such an exponential decay is expected in the early stages, when the container is nearly full, for times less than m_0/c.
(ii) $\Delta = 1$: The amplitude will decay with time in a linear fashion:

$$A(t) = A_0\left(1 - \frac{ct}{m_0}\right). \tag{4.10}$$

(iii) $\Delta = 1/4$:

$$A(t) = A_0 \left(1 - \frac{ct}{m_0}\right)^{\frac{1}{4}}. \tag{4.11}$$

Here the energy loss is dominated by the mass loss from the system.

a is not negligible

In this case (4.3) does not have an analytical solution, but may be solved numerically. An approximate solution may be obtained by assuming that damping is small, so that during one period it has a small influence on the amplitude and the frequency, when compared with A_0 and $\omega_0 \approx \sqrt{(k/m_0)}$, respectively. These assumptions may be expressed by the inequalities

$$b, c \ll \sqrt{km}, \quad a \ll m/A.$$

The amplitude may then be shown to be equal to

$$A(t) \approx \frac{A_0 (1 - ct/m_0)^{\alpha + 1/2}}{1 + \beta [1 - (1 - ct/m_0)^{\alpha}]}, \tag{4.12}$$

where α and β are defined by

$$\alpha = \frac{b}{2c} - \frac{1}{4}, \quad \beta = \frac{16 A_0 \omega_0 a}{3\pi (2b - c)}. \tag{4.13}$$

We can see that in this case A/A_0 does depend on A_0, unlike in (4.7).

Approximate analytic solution for $x(t)$

Based on the functional form in (4.7) one may try a solution for $x(t)$ in the form

$$x(t) = A_0 \left(1 - \frac{ct}{m_0}\right)^{\Delta} \sin(h_0(t) + \varphi). \tag{4.14}$$

When substituted into (4.3) this gives, for $h_0(t)$,

$$h_0(t) = \frac{2\beta}{c} \left[\arctan\left(\sqrt{\frac{1 - \alpha - \tau}{\alpha}}\right) - \left(\sqrt{\frac{1 - \alpha - \tau}{\alpha}}\right) \right], \tag{4.15}$$

where

$$\beta = 1/2 \sqrt{(b + c/2)(b + 3c/2)}, \quad \alpha = \beta^2/m_0 k, \quad \tau = ct/m_0.$$

This solution is valid when the bottle is not near being empty, specifically when the following inequality is satisfied:

$$(b + c/2)(b + 3c/2)/4[m(t)]^2 \ll k/m(t).$$

Goals and possibilities

- Compare the flow rates for a bottle at rest and when oscillating with the spring.
- Find an empirical law for flow rate as a function of outflow diameter by measuring the flow for different diameter openings.
- Measure $A(t)$ and $x(t)$ for different flow rates and characterize the envelope of oscillations as a function of flow rate, both from the point of view of time scales and their curvature (concave, convex). Compare with numerical and approximate analytical predictions.
- How does the period vary with flow rates?
- Introduce into the experimental system a new air resistance factor, by attaching a disk to the bottle, in order to increase the damping during the motion of the bottle being emptied. Careful measurements for a series of flow rates will differentiate between exponential and power law behavior, and between the validity of various models (v or v^2) and solutions.

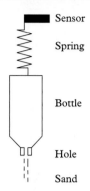

Figure 4.1 *Setup.*

Experimental suggestions

A suggested setup is shown in Fig. 4.1. A spring with a spring constant of around 30 N/m supports a bottle filled with well-sifted uniform-grained sand, say 1.5 kg of it. The latter is the variable mass system. The bottle may be corked with stoppers through which central holes of varying diameters may be drilled, thus allowing for various flow rates. The spring is suspended from a force sensor connected to a data acquisition system. The force sensor allows a direct reading of the flow rate, which can be checked by having the sand gather in a pan resting on an electronic balance.

Sand of various grain sizes may be purchased in order to explore the effect of grain size on the flow. Grain size may be measured under a microscope.

Figure 4.2 shows an arrangement whereby one may check whether the flow rate depends on the acceleration of the bottle, as compared with a stationary bottle. A bucket is suspended from the force sensor, while the bottle is suspended from a fixed stand by a string in one case, and is oscillating on a spring suspended from that stand in the other case. One may analyze the resulting $m(t)$ and dm/dt with an FFT spectral analysis and see whether any correlation exists with the period or frequency of oscillation.

Figure 4.3 shows a wooden disk of 20 cm diameter and 4 mm thickness mounted on the neck of the bottle, thereby increasing the air resistance.

Figure 4.4, reproduced from Ref. [2], shows graphs of $T^2(t)$ for flow rates of 2 g/s (triangles), 15 g/s (squares), and 90 g/s (stars).

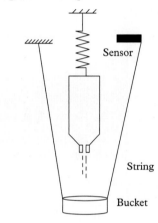

Figure 4.2 *Checking for acceleration.*

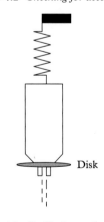

Figure 4.3 *Oscillations damped by disk.*

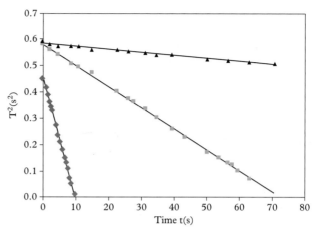

Figure 4.4 *Flow rate dependence of period.*

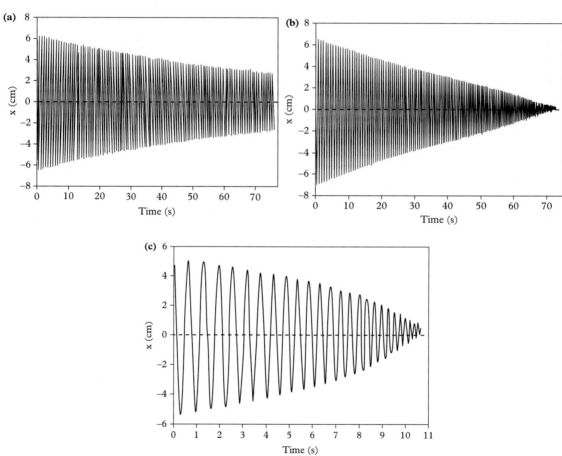

Figure 4.5 *x(t) for three different flow rates: (a) 2 g/s, (b) 15 g/s, (c) 90 g/s.*

Since, by (4.1), m is linear in t, these data confirm the usual formula for the period: $T = \sqrt{(m/k)}$.

Figure 4.5 shows the variation of the position as function of time in a typical experiment, reproduced from Ref. [2], for three types of underdamped oscillations: (a) low flow rate – 2 g/s; (b) medium flow rate – 15 g/s; (c) high flow rate – 90 g/s.

Further explorations

An extension, or a separate project, could be devoted to theoretically treating and experimentally measuring the oscillations of a pendulum suspended from a spring and having a leaking container as the bob – refer to Ref. [4].

....

REFERENCES

1. H.M. Jaeger and S.R. Nagel, Physics of the granular state, *Science* **255**, 1523 (1992).
2. J. Flores, G. Solovey, and S. Gil, Variable mass oscillator, *Am. J. Phys.* **71**, 721(2003), available online: <http://users.df.uba.ar/sgil/web_fisicarecreativa/papers_sg/papers_sgil/Docencia/variabl_mass_osc_2k3.pdf>.
3. R. Digilov, M. Reiner, and Z. Weizman, Damping in a variable mass on a spring pendulum, *Am. J. Phys.* **73**, 901 (2005).
4. N. Danilovic, M. Kovasevic, and V. Babovic, Could a variable mass oscillator exhibit lateral instability? *Kragujevic J. Sci.* **30**, 31 (2008), available online: <http://www.pmf.kg.ac.rs/KJS/volumes/kjs30/kjs30danilovickovacevic31.pdf>.

5 Rotating vertical chain

Introduction 34
Theoretical ideas 35
Goals and possibilities 36
Experimental suggestions 36

Introduction

A cord held horizontally between fixed ends exhibits multiple nodes, separated by half a wavelength, when excited sinusoidally to its various normal mode frequencies. The normal mode frequencies are multiples of the fundamental. However, a rope or chain hanging vertically under its own weight, and therefore with tension varying along its length, when rotated about a vertical axis passing through its equilibrium position and its point of suspension, will exhibit resonances with unequal node spacings at well-defined frequencies, depending on the rate of rotation, as shown in Fig. 5.1. Moreover, these frequencies are not multiples of a fundamental. The project is devoted to the theoretical and experimental exploration of this simple yet unusual system.

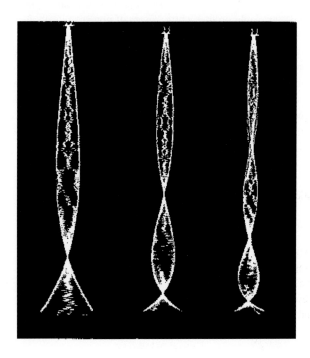

Figure 5.1 *Modes of a rotating vertical chain.*

Theoretical ideas

Two papers [1, 2] have been devoted to the theoretical analysis, and here we summarize the simpler treatment [1].

In the stable configuration of the chain, suspended at $z = L$, its lower end being the point $z = 0$, it oscillates in a plane. Its shape will therefore be described by the function $r(z)$, as shown in Fig. 5.2. Let λ be the linear mass density, T the tension at any point at which the slope of the chain is given by the angle β to the vertical, assumed small. The horizontal force on a small element of mass Δm is given by

$$F_{h,\text{net}} = -\Delta(T \sin \beta) \approx -\Delta(T \tan \beta) = -\Delta(T dr/dz). \tag{5.1}$$

For small slopes, an element of arc length Δs is approximately Δz, so that $\Delta m = \lambda \, \Delta z$. Then the equation of motion is

$$-\Delta(T dr/dz) = (\lambda \, \Delta z)(r\omega^2), \tag{5.2}$$

where ω is the angular frequency of the rotation. The centripetal acceleration $r\omega^2$ may be assumed to be small compared to g (= 9.8 m/s^2). In that case the tension may be assumed to be unaffected by the rotation, and equals λgz. In the limit as Δz goes to zero we find

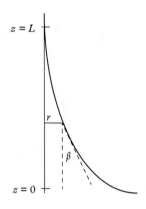

Figure 5.2 *Chain coordinates.*

$$\frac{d}{dz}\left(\lambda gz \frac{dr}{dz}\right) = -r\omega^2, \text{ or} \tag{5.3}$$

$$z\frac{d^2 r}{dz^2} + \frac{dr}{dz} + \frac{r\omega^2}{g} = 0. \tag{5.4}$$

It is convenient to define $x = 2\omega(z/g)^{1/2}$, so that (5.4) transforms into the zeroth-order Bessel equation:

$$x^2 \frac{d^2 r}{dx^2} + x\frac{dr}{dx} + x^2 r = 0. \tag{5.5}$$

Its solution is

$$r = A\mathcal{J}_0(x) + BY_0(x), \tag{5.6}$$

where $\mathcal{J}_0$ and Y_0 are zero-order Bessel functions of the first and second kind. As a first approximation for small oscillations, the end point of the rotating chain remaining close to $z = 0$, we assume it is actually at $z = 0$. Then $Y_0(x)$ must be rejected, since the chain end ($z = 0$) must have a finite solution. We are then left with the stable time-independent solution

$$r = \mathcal{J}_0[2\omega(z/g)^{1/2}]. \tag{5.7}$$

The point of suspension is a node, $r = 0$, so that the normal mode frequencies ω_n for which the chain resonates will be given by $2\omega_n(L/g)^{1/2} = \xi_n$, where ξ_n is the nth zero of the function $\mathcal{J}_0$. For the nth mode ω_n there are n nodes, at positions z_i ($i = 2, 3, \ldots, n$, if we consider only nodes between the lowest one and the point of suspension). We then have $\xi_i = 2\omega_n(z_i/g)^{1/2}$, and the inter-node distances are $(z_i - z_{i-1})$. On squaring and taking the difference between adjacent nodes one finds

$$\xi_i^2 - \xi_{i-1}^2 = (4\omega_n^2/g)(z_i - z_{i-1}). \qquad (5.8)$$

A plot of $\xi_i^2 - \xi_{i-1}^2$ versus $(z_i - z_{i-1})$ yields the normal mode frequencies from the slope, which equals $4\omega_n^2/g$.

We return to (5.6) and (5.7). The shape of the rotating chain will certainly not be that of $\mathcal{J}_0$. But if we take into account that the true endpoint of the rotating chain I is $z > 0$, we may also mix Y_0 into the solution for $r(z)$. The proper admixture A and B of $\mathcal{J}_0$ and Y_0 may then be determined by satisfying the two boundary conditions at the fixed end $z = L$. First, $r(L) = 0$, and secondly, fitting the solution to give the right slope of the chain at $z = L$ (determined either theoretically, by numerically solving the full non-linear differential equation [2], or experimentally, for each mode). This will determine the chain shape $r(z)$ for each mode.

Goals and possibilities

- Determine experimentally as many of the normal mode frequencies for the rotating chain as possible and compare with theory.
- Determine experimentally the exact shape of the chain for each standing wave pattern. In particular, for each mode measure also the slope of the chain at the fixed end. Fit the shape to the theoretical prediction.
- Take three pictures of the spinning chain on the same frame, using a different color filter for each photo. Explore the resulting color addition for different portions along the chain.
- Try chains of different lengths.

Experimental suggestions

The equipment needed includes a variable-speed low-gear motor capable of rotating at slow speeds, from a fraction of a revolution to 4–6 revolutions per second, in steps of a fraction of 1 Hz. Light necklace chains, bathroom plug chains, key ring chains are suitable, say between 30 and 100 cm long. A digital camera or video camera and a frame-grabber program is needed for determining the shape of the rotating chain in the various configurations. Figure 5.1 shows the first three modes [2].

Care needs to be exercised when approaching any resonance from the low frequency side. The chain can be suddenly drawn into very large amplitude patterns by passing slowly through resonance to higher frequencies. These patterns are unstable against perturbations. A check to see that a certain resonance frequency has been exceeded is to straighten the chain briefly: true resonances are self-starting.

The speed of the motor can be measured by a stroboscope, a rotary motion sensor, or a photogate. A stationary background grid of known size is placed behind the chain. This enables one to photograph the chain when it is perpendicular to the line of sight of the camera, and to convert the digital picture during image analysis on the computer into absolute measurements and shapes. One way to analyze the picture is to project the image onto a video monitor and scan the result with a mouse-controlled cursor to record a large number of coordinates along the chain image. A similar scan of the background grid allows one to convert the picture into real space coordinates.

REFERENCES

1. A.B. Western, Demonstration for observing $\mathcal{J}_0(x)$ on a resonant rotating vertical chain, *Am. J. Phys.* **48**, 54 (1980).
2. M.P. Silverman, W. Strange, and T.C. Lipscombe, "String theory": Equilibrium configurations of a helicoseir, *Eur. J. Phys.* **19**, 379 (1998).

6

Cycloidal paths

Introduction	38
Theoretical ideas	38
Goals and possibilities	40
Experimental suggestions	40

Introduction

Let a particle move from point A to a lower point B in a constant gravitational field. Find the equation of the track from A to B along which the time for travel will be a minimum. This is the brachistochrone problem proposed by Johannes Bernoulli in 1696. The required equation of the track turns out to be a cycloid. In addition, the period of oscillatory motion about the lowest point of the track for a particle released from any point above it is independent of the initial position, a property known as the tautochrone. The project involves building cycloidal and inclined plane tracks, confirming experimentally the tautochrone property of the former and comparing the times of descent between identical-height drops on the two tracks. In addition a model is built which illustrates the principle of the first accurate pendulum clock constructed by Huygens.

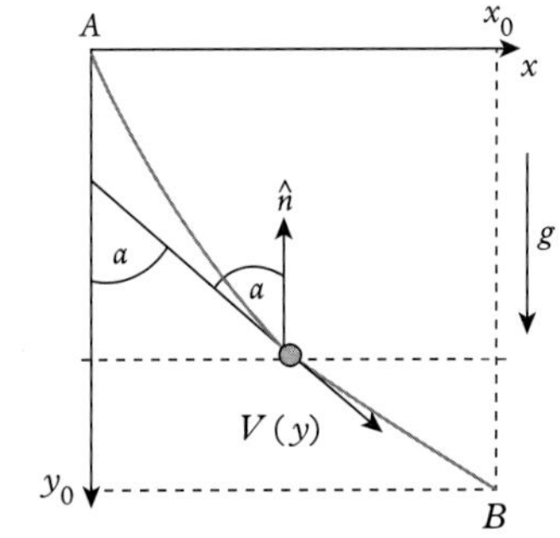

Theoretical ideas

See Ref. [1]. The problem is presented in Fig. 6.1: the particle is to slide from point $A(0,0)$ to point $B(x_0,y_0)$ in minimum time. It is often solved by using the calculus of variations. We present here a solution using Fermat's theorem of minimum time in optics. If the particle followed the path of a ray of light it would surely cover the distance in a minimum time. How do we map the ray optics to the particle motion problem?

The velocity of the particle along the track varies, whereas the velocity of light is constant in a homogeneous medium. But suppose we divide the space between A and B into horizontal parallel slices, as in Fig. 6.2, such that the optical density is highest at A (velocity V lowest) and lowest at B, in which case the ray will follow a path curved in just the right way. Snell's law, (an example of Fermat's minimum time principle), gives $(\sin \alpha_i)/V_i$ = const. In the limit of infinitesimally thin layers, α and V will be continuously varying quantities depending on x and y, and one has

$$(\sin \alpha_{x,y})/V_{x,y} = \text{const.} \qquad (6.1)$$

For the material particle, $V = \sqrt{(2gy)}$. Suppose we choose the lowest point at $y = 2R$ (the meaning of R will be clarified presently), where $\alpha = 90°$. At that

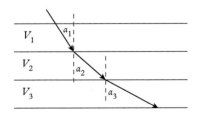

Figure 6.1 *Coordinates.*

Figure 6.2 *Snell analogue.*

point $V = \sqrt{(2gy)} = \sqrt{(4gR)}$, and therefore the constant in (6.1) is $1/\sqrt{(4gR)}$, so that (6.1) may be rewritten as

$$V_{x,y} = \sin \alpha_{x,y} \sqrt{(4gR)}. \qquad (6.2)$$

This is the velocity of the material particle along the path where the light ray would travel, and is therefore covered in the shortest time.

The path is obtained by integrating the velocity, for which we must know $\alpha_{x,y}$ – see Fig. 6.3. It may be obtained from kinematic considerations. On the one hand, the acceleration at any point is, from (6.2),

$$a = dV/dt = (d\alpha/dt)\cos\alpha \sqrt{(4gR)}. \qquad (6.3)$$

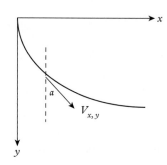

Figure 6.3 $V_{x,y}(\alpha_{x,y})$.

On the other hand, at any point the particle is instantaneously on an inclined plane along the tangent to the curve, Fig. 6.4, having an acceleration $g\sin\theta = g\cos\alpha$. Comparing with (6.3) we get

$$(d\alpha/dt) = \sqrt{(g/4R)} \equiv b \text{ (constant)}, \qquad (6.4)$$

so that α grows linearly with time, namely $\alpha = bt$. Substituting back into (6.2) we have

$$V = (4R)b\sin(bt). \qquad (6.5)$$

It is easiest to integrate this by resolving into components. Thus we have

$$V_x = V\sin\alpha = 4Rb\sin^2(bt) = 2Rb(1-\cos 2bt),$$
$$V_y = V\cos\alpha = 2Rb\sin(2bt).$$

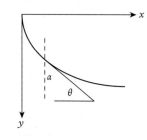

Figure 6.4 *Inclined plane.*

Performing the integrations and rearranging,

$$\begin{aligned} x &= R(2\alpha - \sin 2\alpha), \\ y &= R(1 - \cos 2\alpha). \end{aligned} \qquad (6.6)$$

These are the parametric equations of a cycloid, our final result for the brachistochrone, which is the locus of a point on a rolling hoop of radius R. The starting point is $(0,0)$, the lowest point is $(\pi R, 2R)$, as shown in Fig. 6.5. Given any two points $A(0,0)$ and $B(x_0,y_0)$, the radius of the appropriate cycloid passing through them is given by the (numerical) solution of (6.6).

Expression (6.5) is just like velocity in simple harmonic motion, so that the period of motion of the particle is

$$T = 2\pi/b = 2\pi\sqrt{(4R/g)}. \qquad (6.7)$$

It is independent of the amplitude: whatever the point of release A, the particle will arrive at the same time at the bottom – the tautochrone property. The equivalent length of a simple pendulum is $4R$.

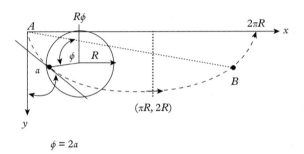

Figure 6.5 *Cycloid parameters.*

The regular pendulum is, of course, not isochronous: the period does depend on the amplitude. But it may be shown that the period of a mathematical pendulum of length $4R$ will be exactly independent of its amplitude, provided that it is suspended at the meeting point of two adjacent cycloids generated by a wheel of radius R, and its suspension is confined to move along these cycloids as it swings from side to side, as shown in Figs 6.7 and 6.8. This is the principle on which Huygens built the first accurate pendulum clock. The bob itself describes a cycloid.

The extension from a particle to a ball rolling without slipping in the minimum time is simple. Instead of $V = \sqrt{(2gy)}$, the moment of inertia $I = kmr^2$ ($k = 0.4$ for a ball) also enters into energy conservation, $mgy = mv^2/2 + I\omega^2/2$, so that $V = \sqrt{[2gy/(1+k)]} = \sqrt{(2g^*y)}$, where $g^* = g/(1+k)$. The rest is as before, with $g \to g^*$ everywhere.

Goals and possibilities

- Build a cycloidal track and for comparison purposes also a straight, variable-angle inclined track.
- Measure the time taken for a ball to roll from various release points on the brachistochrone to the bottom of the track. Compare with the theoretical formula. Compare the time taken to descend the same height on the brachistochrone and on the straight track.
- Design a pendulum whose bob is constrained to move along a cycloid, and whose suspension is confined by cycloids on either side of its swing from the equilibrium position. Measure its period for various amplitudes and show that it is a constant.

Experimental suggestions

Mark a point on the circumference of a hoop, lid, or other circular object, whose radius you have measured. Roll it in a vertical plane and trace the locus of the point on a piece of cardboard placed behind the rolling object. Transfer the trace to a 2 cm-thick board and cut out with a jigsaw.

The cycloid will be the line between the upper and lower parts in Fig. 6.6. Lay along the profile line a flexible plastic track, obtainable from household or electrical supplies stores. The cycloidal track is ready for experimentation.

Figure 6.7 shows the brachistochrone, the wheel which served to generate it, one of two photogates for measuring the period, and the variable-angle straight inclined track. Saw the upper part of Fig. 6.6 accurately into two along a perpendicular line from its lowest point, and place the two pieces side by side, as shown in Fig. 6.8.

Figure 6.6 *Cutting a cycloid.*

Suspend a simple pendulum of length $4R$ at the apex of the cusp, as shown in Fig. 6.9. The bob itself will describe a cycloidal path during its oscillation, its suspension is limited on either side by cycloidal tracks, and therefore its period will be independent of amplitude. The measurements of the period, both of the rolling ball and of the cycloidal pendulum, may be carried out with one or more photogates.

Figure 6.10 shows the period as a function of the passage number for a steel ball released from the top of the brachistochrone and left to oscillate about the center [2]. The period is the time elapsed between two successive intersections of the photogate beam on either side of the center. There is a gradual small decrease

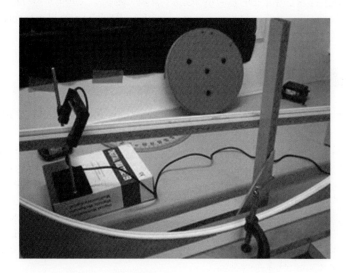

Figure 6.7 *Brachistochrone and incline.*

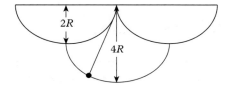

Figure 6.8 *Creating the cusp.*

Figure 6.9 *Cycloidal pendulum.*

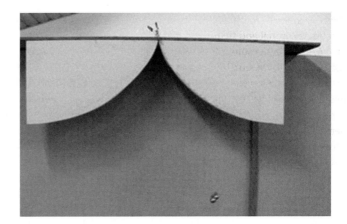

Figure 6.10 *Period on brachistochrone.*

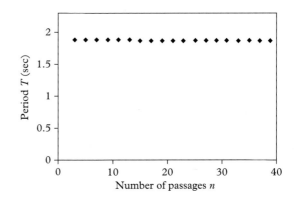

Figure 6.11 *Period on cycloidal pendulum.*

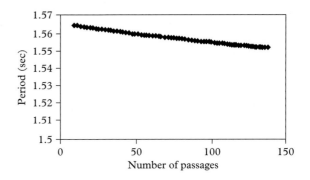

in the period, instead of it being a constant, due to friction and irregularities of the track, but the initial and final periods differ only by 1.5%.

Figure 6.11 shows results [2] of the period versus amplitude for the model Huygens pendulum, the bob having been released from a position where its suspension completely covered one cycloid track on one side of its swing. There is a gradual decrease in the period, but the difference between the initial and final times is only 0.76%. This is well outside the amplitude variation of the period for a mathematical pendulum, for which the period departs from the small amplitude value (for which $\sin\theta = \theta$) by anything between 2% and 18% as the angular amplitude varies from 30° to 90°.

REFERENCES

1. D. Agmon and P. Gluck, *Classical and Relativistic Mechanics* (World Scientific, Singapore, 2009).
2. P. Gluck, Motion on cycloid paths: A project, *Phys. Educ.* **45**, 270 (2010).

7 Physics of rubber bands and cords

Introduction 44
Theoretical ideas 44
Goals and possibilities 45
Experimental suggestions 46

Introduction

Rubber bands and cords, though elastic, behave remarkably differently from helical springs or the inextensible stretched strings of a musical instrument. For small stresses their stress–strain curve is non-linear. For larger stresses, in a complete cycle of loading and unloading, their stress–strain curve exhibits hysteresis. Plucking a rubber band while gradually stretching it and thereby increasing the tension produces a practically unchanging pitch. The pitch does change if the length is constant but the tension increases. The amount of light transmitted through a thin flat rubber band varies with the tension in unexpected ways. For some types of rubber, heating may cause contraction rather than expansion. These phenomena are sufficiently unusual to investigate quantitatively in an experimental project.

Theoretical ideas

An empirical equation of state of a rubber cord for small loading, no deformation, and no hysteresis is given by [1,2]

$$F = AT(L/L_0 - L_0^2/L^2), \qquad (7.1)$$

where L_0 is the unstretched length, L is the length of the band when a force F is applied, T is the absolute temperature, and A is a constant.

Suppose a body of mass m is attached to one end of a vertically hanging rubber cord and allowed to fall from a position in which the strain $\varepsilon (= \Delta L/L)$ has the value ε_0. There are two aspects to consider. First, for the rubber cord Hooke's law must be modified in order to account for the time dependence of the strain. As an approximation one might assume a relation [3]

$$\sigma = Y\varepsilon + \eta\dot{\varepsilon}, \qquad (7.2)$$

where σ is the stress, Y is Young's modulus and η is some internal "viscosity" of the material (though Y and η might themselves be a function of ε).

Secondly, if (7.2) is valid, the resulting motion of the mass will be governed by

$$S(Y\varepsilon + \eta\dot{\varepsilon}) = m(g - L\ddot{\varepsilon}), \qquad (7.3)$$

where S is the cross-sectional area of the cord. This equation is akin to that of damped simple harmonic motion, and the analysis is similar. If, as is usually the case, $4YmL > S\eta^2$, the solution to (7.3) is

$$\varepsilon = \frac{mg}{SY} - \left(\frac{mg}{SY} - \varepsilon_0\right) e^{-Kt} \cos\omega t, \qquad (7.4)$$

where $K = S\eta/2mL$, ε_0 is the initial strain of the cord at which the mass begins its fall, and $\omega = \sqrt{(S/mL)[Y - (S\eta^2/4mL)]}$. If K and ω are measured from dynamic experiments, the values of Y and η may be determined from $Y = mL(K^2 + \omega^2)/S$, and $\eta = 2mLK/S$.

A good source for the physical properties of rubber is an old but very readable book by Treloar [4]. A more advanced and up-to-date text is Ref. [5]. You should also watch the valuable online videos [6–8], among them Richard Feynman talking about the physics of rubber.

Goals and possibilities

Mechanical properties

Measure the extension of a rubber band while loading and unloading. Do this for very small loads where hysteresis is negligible, say up to 150 g, and try to fit results to the equation of state relating the force to the extension. A rubber band of thickness 1 mm and length say 6–10 cm is useful for this purpose. For larger loads study the time scale over which the rubber cord continues to stretch before reaching the final strain, as well as the stress–strain curve for loading and unloading, in order to draw the hysteresis curve and from it the energy loss in a cycle. Show that the elastic modulus is a function of strain.

Set the loaded rubber band into small vertical oscillations for various loads in order to determine the dynamic "spring" constants and compare with the static values. If the oscillation frequency is great enough, so that relaxation cannot be completed in an oscillation cycle, the dynamic moduli will differ from the static ones. Compare these in a graph of static and dynamic moduli versus strain. Try to extract Y and η from the experiment.

Acoustic properties

Pluck the rubber band in two different experimental setups, varying the tension in both cases. In the first case the rubber strip is held horizontally and is stretched by passing it over a pulley and loading it with various weights. One has a constant length here, from the fixed end to the pulley, and the pitch varies with tension. In the second case the band is held vertically: fixed at the top end to a force sensor which measures the tension, and to a movable clamp attached to a lab stand at the other, whose position determines the tension. When plucked, the pitch produced is recorded with a microphone as a function of length and tension. Results can be compared with the variation of frequency in a (nearly inextensible) string of a musical instrument as the peg increases the tension.

Transmission of light

Measure the amount of light transmitted through the strip as a function of strain and try to correlate the results with findings of the mechanical and acoustic properties [3]. Polarization and birefringence properties of rubber strips under varying strain could be investigated, extending the project very considerably [3].

Thermodynamics

Let $f(F,L,T)$ be the function of state for the rubber band, where F, L, and T are the tension, length, and temperature of the rubber, respectively. Then thermodynamics shows the following relation between partial derivatives:

$$(\partial F/\partial L)_T (\partial L/\partial T)_F (\partial T/\partial F)_L = -1.$$

Devise a series of experiments in which all these derivatives may be measured and the above relationship may be verified.

Experimental suggestions

There are many types of rubber bands and strips which can be used: from hosiery (peel away the external cloth wrapping to reveal thin rubber strands of circular cross section), model airplane wind-up motors, old inner tubes, balloons, commercial rubber bands, and so on. A flat, natural-color, rectangular-profile strip some 8 mm wide unstretched and of thickness say 0.6 mm is suitable for light transmission studies. Rubber cords and bands of anything between 0.6 mm and 10 mm thick and of varying lengths, to match the experimental apparatus designed, are suitable for mechanical and acoustic measurements.

For the mechanical part, both the stress–strain relationship and the oscillations, suspending the strip from a force sensor and measuring the descent of the weights attached at the other end with a sonic ranger will give an accurate and quick computer-assisted recording of the results. After each weight, both in loading and unloading, wait at least two minutes for relaxation to take place before measuring the strain. Since the relaxation frequency of the rubber is much less than 1 Hz, the oscillations of a mass suspended from the strip can be of order 1–2 Hz.

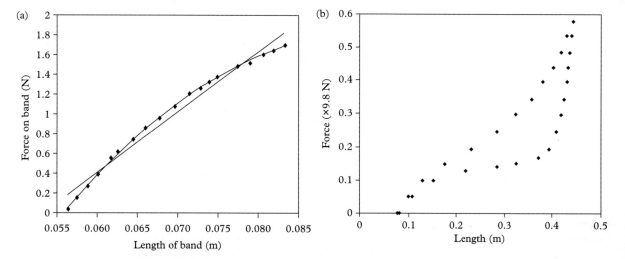

Figure 7.1 *(a) Stress–strain curve for small loads. (b) Stress–strain curve for large loads.*

Very large weights could be used, resulting in low frequency oscillations and attendant relaxation effects. Sensitive thermocouple sensors could then possibly monitor the attendant temperature changes.

Typical stress–strain graphs are shown in Fig. 7.1. In Fig. 7.1(a), a rubber band 1 mm thick and 5.5 cm long was loaded in steps of 5 g up to 150 g. The straight line graph is a least squares linear fit to Hooke's law, clearly not representing well the experimental data [9]. The data fit better a quadratic curve, as predicted by (7.1). Figure 7.1(b) shows hysteresis for much larger loads.

In the light transmission experiment use a powerful LED as the light source and a photodiode as the detector. In order to eliminate background light encase the rubber strip in a channel, carved in a piece of wood, which allows free movement of the strip but is light proof, as shown in Fig. 7.2. On one side of the enclosure embed an LED, and on the other side a photodiode. A graph of photodiode current versus stretch for one rubber band is shown in Fig. 7.3. One sees that initially the strip actually gets optically denser as it is stretched, even though it is getting thinner. The phenomenon could be due to Poisson's ratio effects, as well as to increased light scattering, resulting in the rubber becoming more opaque, a behavior also found when one bends and distorts hard plastics. Make sure that the measurements range over a strain of some 200%.

In order to study $(\partial F/\partial T)_L$, the setup in Fig. 7.4 may be used. Attach to the band a 200 g mass and place the latter on a balance. Set the balance off equilibrium, then the stretching force exerted by the mass is shared by the band and the balance, thereby enabling one to adjust the force the mass exerts on the band. Choose a force, zero the balance, measure the band length and the temperature (by a small thermocouple positioned close to the band). Heat the band and measure the final temperature and the change in the force due to the

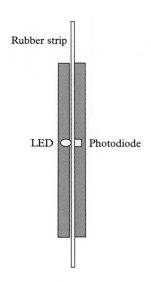

Figure 7.2 *Light transmission setup.*

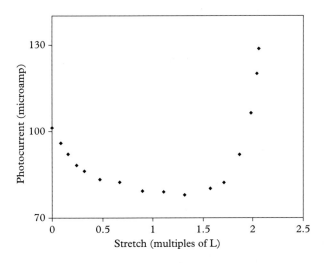

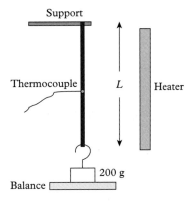

Figure 7.3 *Stretch dependence of photocurrent.*

Figure 7.4 *To study $(\partial F/\partial T)_L$.*

band's tendency to contract on heating. For uniform heating of the band use a space heater. Use a vernier caliper or a traveling microscope to measure the length.

In the acoustic part use a microphone to record the sound from the plucked band, connected to a data acquisition system, in order to determine the frequency and harmonic content of the sounds produced.

....

REFERENCES

1. L.E. Reichl, *A Modern Course in Statistical Physics* (University of Texas Press, Austin, 1980), p. 56.
2. T.L. Hill, *An Introduction to Statistical Thermodynamics* (Dover, New York, 1986), chapters 13 and 21.
3. T.B. Brown, *The Taylor Manual* (Addison-Wesley, Reading, MA, 1959), p. 37.
4. L.R.G. Treloar, *The Physics of Rubber Elasticity* (Oxford University Press, Oxford, 1949).
5. G. Strobl, *The Physics of Polymers* (Springer, Berlin, 2007).
6. R. Feynman, *Rubber Bands*, <http://www.FeynmanPhysicsLectures.com/fun-to-imagine/rubber-bands>.
7. R. Allain, *Do rubber bands act like springs?* Wired.com, 2012. Available at <http://www.wired.com/2012/08/do-rubber-bands-act-like-springs/>.
8. A. Family, *Hysteresis and rubber bands*, 2006. Available at <http://www.madphysics.com/exp/hysteresis_and_rubber_bands.htm>.
9. G. Savarino and M.R. Fisch, A general physics laboratory investigation of the thermodynamics of a rubber band, *Am. J. Phys.* **59**, 141 (1991).

Oscillation modes of a rod

Introduction

A cylindrical rod suitably prepared may be made to oscillate around many points and axes, thereby revealing a surprising variety of physical phenomena, ranging from a regular physical pendulum through bifilar twisting motion and torsional oscillations, all the way to possible chaotic behavior in a double pendulum [1–3]. The construction of the various suspension and support points, the period measuring device, and the theoretical calculations of the appropriate periods all require ingenuity and increasing depths of understanding of the rigid body dynamics involved, and provide plenty of scope for playing around with the physics.

Introduction	49
Theoretical ideas	49
Experimental suggestions	52

Theoretical ideas

Let us suppose that we have prepared some means of supporting the rod in a series of positions illustrated by the diagrams that follow, Figs 8.1–8.12. In each case we shall describe the modes of oscillation and provide the formula for its period – proving them is in some cases a non-trivial exercise and may be considered part of the project. Proofs are given in the more complex cases.

Figure 8.1 shows a rod of radius R, length L, and mass M having a series of holes drilled symmetrically on either side of the center of mass (*com*). The holes are of radius r_h. A pin of radius $r_p < r_h$ can be used to suspend the rod from any of the holes, or strings can be attached to the rod through them.

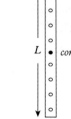

Figure 8.1 *Drilled rod.*

Vertical pendulum

The rod can oscillate in a vertical plane by suspending it from a pin of radius r_p through one of the holes, as in Fig. 8.2. Figure 8.3 shows that the top of the hole will roll on the pin without slipping, keeping the point of contact at rest. If the distance from the *com* to the top of the hole is d, and if one neglects the finite diameter of the pin, the period for small oscillations is given by

$$T_0 = 2\pi \sqrt{\frac{(L^2/12) + d^2}{gd}}. \tag{8.1}$$

T as a function of d will exhibit a well-known minimum.

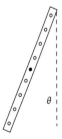

Figure 8.2 *Vertical oscillations.*

Figure 8.3 *Finite radius.*

A good exercise is to extend this formula in two ways: First, to large amplitudes, in which case the period will be given by an elliptic integral [1] which can be evaluated numerically for any amplitude; secondly, to take into account the finite radius of the pin, shown in Fig. 8.3.

Horizontal pendulum

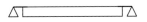

Figure 8.4 *Horizontal pendulum.*

Pins, attached to the ends of the rod, are supported horizontally, as shown in Fig. 8.4. The rod oscillates about an axis parallel to its long axis. Its period is given by

$$T = 2\pi\sqrt{3R/2g}. \tag{8.2}$$

Bifilar suspension

Here, the rod is suspended horizontally by two vertical parallel strings which may be of equal length H, as in Fig. 8.5 (H is measured from the point of suspension to the rod axis), or of different lengths, as in Fig. 8.7.

Consider first equal length strings. Several motions are possible.

(a) In and out of its equilibrium plane, like a swing. The axis of rotation is parallel to the rod, passing through the suspension points of the strings. The period is given by

$$T = 2\pi\sqrt{(0.5R^2 + H^2)/gH}. \tag{8.3}$$

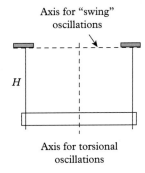

Figure 8.5 *Bifilar suspension.*

(b) Laterally, parallel to the rod length, in its equilibrium plane. The period is given by

$$T = 2\pi\sqrt{(H-R)/g}. \tag{8.4}$$

(c) Torsional oscillations about a vertical axis through the center of mass (see Fig. 8.5). In this case, while the left end of the rod moves out of the paper the right end moves into the paper, and vice versa. The period for small oscillations will be

$$T = 2\pi\sqrt{\frac{H'}{g}\frac{L/d}{\sqrt{12}}}, \tag{8.5}$$

where $H' = H - R$.

For large amplitude oscillations the period will once again be given by an elliptic integral, to be evaluated numerically.

(d) Inclined bifilar pendulum. Let the system execute torsional oscillations about a vertical axis, making an angle θ with the rod at the *com* of the rod – see Fig. 8.6. The moment of inertia about the vertical axis is given by

$$I = M\{L^2/12 + R^2/4\}\cos^2\theta + M\{R^2/2\}\sin^2\theta. \tag{8.6}$$

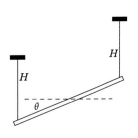

Figure 8.6 *Inclined pendulum.*

For small angles, say $\theta < \pi/6$, the period is expected to be the result of (8.5) multiplied by $\cos\theta$.

For swing-like oscillations, the result of (8.3) is still valid.

(e) **Horizontal rod suspended from unequal-length strings.** Let d be the distance of the points of attachments of the strings to the center of mass, as in Fig. 8.7. In a torsional displacement of the rod the attachment points of the string move out of the equilibrium plane by x_1 and x_2, the center of mass moves by x_{cm}, while the rod swings by an angle θ, as in Fig. 8.8. These parameters are related by $x_{cm} = (x_1 + x_2)/2$ and $\theta = (x_1 - x_2)/2d$.

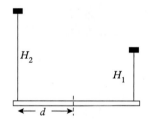

Figure 8.7 *Unequal suspensions.*

The kinetic and potential energies of the system in terms of these parameters are given by

$$E_k = \frac{1}{2}I_{cm}\dot\theta^2 + \frac{1}{2}M\dot x_{cm}^2, \tag{8.7}$$

$$U = Mgy_{cm} = \frac{1}{2}Mg\left(H_1 + H_2 - \sqrt{H_1^2 - x_1^2} - \sqrt{H_2^2 - x_2^2}\right). \tag{8.8}$$

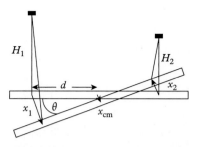

The Lagrangian is $L = E_k - U$. In the small-angle approximation one obtains the coupled Euler–Lagrange equations for the two variables θ and x_{cm}:

$$\ddot x_{cm} = -\omega_0^2(x_{cm} - \varepsilon d\theta), \tag{8.9}$$

$$\ddot\theta = -\omega_0^2\lambda(\theta - \varepsilon x_{cm}/d), \tag{8.10}$$

Figure 8.8 *Coordinates for unequal suspensions.*

in which

$$\omega_0^2 = g/H_{eff},\ \lambda = md^2/I_{cm},\ \varepsilon = (H_1 - H_2)/2H_{eff},\ H_{eff} = 2H_1H_2/(H_1+H_2).$$

From (8.9) and (8.10) one gets the normal mode frequencies

$$\omega_\pm^2 = \omega_0^2\left\{\frac{1}{2}(1+\lambda) \pm \sqrt{4(1-\lambda)^2 + \lambda\varepsilon^2}\right\} \tag{8.11}$$

and the periods $T_\pm = 2\pi/\omega_\pm$.

Double pendulum

Here the rod is hung from a string passing through one of its holes, as in Fig. 8.9. String and rod can either swing together or in opposite directions. The method of deriving the normal mode frequencies is similar to that for case (e) above; this time the dynamical variables are θ_1 and θ_2, and the frequencies are given by

$$\omega_\pm^2 = \frac{\omega_0^2}{2(1-d/s)}\left\{(H/s+1) \pm \sqrt{(H/s+1)^2 - 4(H/s)(1-d/s)}\right\}, \tag{8.12}$$

where $s = [L^2/12 + d^2]/d$, $\omega_0^2 = g/H$.

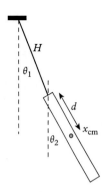

Figure 8.9 *Double pendulum.*

Chaotic pendulum

As shown in Fig. 8.10, a steel wire passing through a hole near the *com* and suspended by very short strings to a support allows the rod to swing freely about its axis of support. Releasing the upper part of the pendulum from a large angle will result in chaotic oscillations of the rod, swinging unpredictably in one direction or the other. Numerical solution of the Euler–Lagrange equation for the system will show extreme sensitivity to the initial conditions.

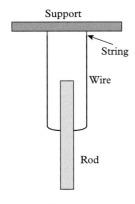

Figure 8.10 *Chaotic pendulum.*

Torsional pendulum

The torque in a wire of length H and radius r when twisted through an angle θ is given by

$$\tau = -\frac{G\pi r^4}{2H}\theta, \tag{8.13}$$

where G is the shear modulus. The period is the well-known expression

$$T = 2\pi\sqrt{2HI/G\pi r^4}. \tag{8.14}$$

The moment of inertia I depends on whether the rod is suspended horizontally or vertically, shown for both cases in Fig. 8.11.

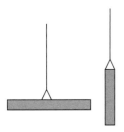

Figure 8.11 *Torsional pendulums.*

Horizontal pendulum

Suspend the rod at one end by a string, while a second point on the rod, P, rests on a support which is a horizontal distance d from the *com*, and a distance s from the point of attachment of the string, as in Fig. 8.12. Let the rod swing through an angle φ about P, then, out of the plane, the string moves through an angle θ such that $H\theta = s\varphi$. In the small-angle approximation, the torque exerted by the string is given by $\tau = F\theta s = Fs^2\varphi/H$, where the tension in the string is $F = Mgd/s$. Hence the period is given by

$$T = 2\pi\sqrt{\frac{HI}{Mgds}} = 2\pi\sqrt{\frac{H(L^2/12 + d^2)}{gds}}. \tag{8.15}$$

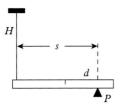

Figure 8.12 *Horizontal pendulum.*

Experimental suggestions

- Periods can be timed manually, but preferably with suitably positioned photogates.
- For the vertical pendulum, for large angle oscillations it is best to use the outermost hole in the rod.

- In case (e) for bifilar suspension, release the rod after rotating it through a small angle about a string.
- For the double pendulum, the normal modes are obtained by starting with the string and rod displaced in the same direction, or in the opposite direction.
- Experiment with different suspension lengths.
- Invent oscillations that have not been described here; many variants are possible.
- If you have a device that can follow and record two- or three-dimensional motion, such as Alberti's windows [4] (which uses two coupled video cameras), the displacement versus time of the pendulums may be sampled and compared with numerical solutions of the differential equations of motion.

REFERENCES

1. A. Cromer, Many oscillations of a rigid rod, *Am. J. Phys.* **63**, 112 (1995).
2. R.A. Nelson and M.G. Olson, The pendulum: Rich physics from a simple system, *Am. J. Phys.* **54**, 112 (1986).
3. A. Cromer, C. Zahopoulos, and M. Silevich, Chaos in the corridor, *Phys. Teach.* **30**, 382 (1992).
4. Alberti's Window, *Alberti's window motion visualizer*, n.d. <http://www.albertiswindow.com>.

Part 2

Electromagnetism

9	Physics of incandescent lamps	57
10	Propulsion with a solenoid	62
11	Magnetic dipoles	69
12	The jumping ring of Elihu Thomson	74
13	Microwaves in dielectrics I	80
14	Microwaves in dielectrics II	86
15	The Doppler effect	89
16	Noise	94
17	Johnson noise	102
18	Network analogue for lattice dynamics	106
19	Resistance networks	111

Physics of incandescent lamps

9

Introduction	57
Theoretical ideas	57
Goals and possibilities	59
Experimental suggestions	59

Introduction

A great deal of physics may be learnt by investigating the thermal, electrical, and optical properties of incandescent lamps consisting of a tungsten filament in a glass envelope or bulb filled with inert gas at low pressure: conduction, convection, and radiation may all play a part in normal operation, as does the connection between temperature, resistance, and the spectrum emitted. The spectrum of radiation produced at various temperatures is relevant to the Wien displacement, Stefan–Boltzmann, and Planck laws. The burning out of a lamp and its lifetime depends on the operating voltage, and it is of interest to model that behavior. Moreover, when operated with AC power at various frequencies, hysteresis effects appear. A variety of these effects are the subject of this chapter.

Theoretical ideas

The background and formulae for the Wien, Stefan–Boltzmann, and Planck radiation laws may be found in standard texts. Intially, we shall restrict attention to outlining a mechanism for burnout lifetime [1] as a function of applied external voltage V. Three quantities play key roles in determining the lifetime of the tungsten filament, namely, resistivity, emissivity, and rate of evaporation of the filament material. All these are temperature dependent.

Consider first the temperature-dependent resistance. When a constant (DC) voltage V is applied to a filament of resistance R the electric power dissipated in the filament is

$$P_{el} = IV = I^2 R = V^2/R. \qquad (9.1)$$

The associated energy may be dissipated through several mechanisms: heat conduction and thermal convection to the bulb and surroundings through the filling gas, thermal conduction through the leads, and radiation. A small fraction of the latter is visible, the rest is mostly in the infrared. For a uniform wire at any temperature T (in kelvin) the resistance in (9.1) is given by

$$R(T) = \rho(T)\ell/a, \qquad (9.2)$$

where ρ, ℓ, a are the resistivity, length, and cross-sectional area of the wire, respectively. For pure tungsten the following formula fits the experimental data well [2]:

$$\rho(T) = 5.792 \times 10^{-11} T^{1.209}. \qquad (9.3)$$

Thus, the filament temperature may be determined from (9.3) in terms of $R(T)$ and the room temperature resistance $R(T_0)$ (when no voltage is applied):

$$T = T_0 [R(T)/R(T_0)]^{0.8271}. \qquad (9.4)$$

Secondly, consider the power *radiated* by the filament. This involves a material- and T-dependent emissivity $e(T)$ and the effective radiating surface area A_{sfc}:

$$P_{\text{rad}} = \sigma e(T) A_{\text{sfc}} (T^4 - T_0^4), \qquad (9.5)$$

where one may safely neglect T_0^4, since $T/T_0 \approx 10$. Here $\sigma = 5.67 \times 10^{-8} \text{W}/(\text{m}^2\text{K}^4)$ is the Stefan–Boltzmann constant, and the emissivity is dimensionless and equal to unity for a perfect emitter. For tungsten, an *empirical* formula for the emissivity is [2]:

$$e(T) = 1.731 \times 10^{-3} T^{0.6632}. \qquad (9.6)$$

A third material property of the filament is its evaporation rate $E(T)$, in kg/(s·m^2). Published data for tungsten conforms to a power law $E(T) = MT^N$, with considerable variability in the values of M and N. For instance, N, which determines the power of the voltage dependence of the filament life, varies between 23 and 49, depending on the data chosen and the temperature range. In the range 3000 – 3700 K, one set of data for tungsten is well represented by the expression [2]:

$$E(T) = 7.977 \times 10^{-137} T^{37.76}. \qquad (9.7)$$

It is interesting to consider now the lifetime of a bulb (burnout) as a function of external voltage. This may be modeled by assuming that there is a so-called hot spot, a very short length of the filament that is thinner (of higher resistance) than the rest. In mass production of lamps this is sure to occur. As the tungsten evaporates faster there, this only exacerbates the situation locally: at the hot spot, temperature, rate of evaporation, and shrinkage all *accelerate*, leading to melting and eventual rupture. A detailed calculation for such a model has been performed [1]; it considers the local temperature, evaporation rate, and shrinkage rate of a hot spot, assuming that the electrical power dissipated is approximately equal to the power radiated in (9.5). In this (simplified) model, the lifetime of the filament was shown to depend on external voltage as $V^{-12.86}$. Manufacturers' data [3–5] on burnout also indicate a V^{-n} dependence, where $10 < n < 16$.

Goals and possibilities

- Measure the I–V curve and from it determine the resistance of the lamp as a function of V. Then use (9.3) to find the temperature of the filament as a function of V.
- Use a thermopile to measure the radiation emitted by the lamp at different temperatures (voltages). Notice that the intensity (power per unit area) of emitted radiation is given by (9.5). Use (9.5) and (9.6) to check whether the Stefan–Boltzmann law holds. Note: the spectral response of a thermopile is essentially flat in the infrared region (from 0.5 to 40 µm): *why does this matter?*
- Used in conjunction with a spectrometer, one might attempt to map out the spectral distribution of the radation emitted by the lab at different termperatures (voltages).
- Alternatively, use a small calorimeter suitably insulated to measure the total heat output of the lamp.
- Connect the lamp in series with a charged capacitor and follow its discharge. The behavior will differ from the standard one since the resistance is variable.
- Measure the thermal response of the lamp: how soon after connecting to the power source does it reach the operating current? Similarly, after disconnecting, what is the rate at which the filament cools down?
- Investigate how repeated current surges, occurring when switching on and off at some preset fast rate, influence the lifetime of the lamp.
- Measure the lifetime t_L of the lamp as a function of the applied voltage.
- Interesting hysteresis effects arise when the lamp is operated at very low frequency AC voltages, in the region of 0.01, 0.1, and 1 Hz. Plot the I–V characteristics under these conditions to show the hysteresis.

Experimental suggestions

Use *cheap*, small incandescent lamps such as Christmas lights, auto 12 V lights, or flashlight bulbs, which operate at low voltages and currents, with no inert gas inside, making convection and conduction negligible.

The apparatus required for this project consists of digital multimeters; function generator; spectrometer; computer-controlled current, voltage, and light sensors capable of high sampling rates. It *is* important that the detector response have very little dependence upon wavelength, and for this reason a thermopile is used (which has a nearly wavelength-independent response over the range from 400–5000 nm). A thermopile is composed of multiple thermocouples measuring the temperature of a highly absorbing surface; it outputs a voltage that is proportional to the influx of power.

Lifetime

Use a batch of 100 or more nominally identical, inexpensive lamps. Operate each group of 10 at some voltage higher than the recommended one, until burn-out. Measure lifetime by continuous computer-operated logging of lamp current, which drops to zero when the filament breaks. Calculate the lifetime for each voltage from an average for the 10.

Thermal response

Warmup

Rather than use an ammeter, simply connect the lamp in *series* with a 0.1 Ω precision resistor R (rated to handle the expected power) and a stabilized power supply V_s. Measure the voltage across the resistor $V_R(t)$ as a function of time at a sampling rate of 20 kHz. This gives the current $I(t)$. The resistance of the lamp is then $R(t) = [V_s - V_R(t)]/I(t)$. Using (9.4) we can then calculate $T(t)$.

Cooling

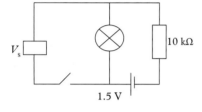

Figure 9.1 *Circuit for cooling.*

Use the *parallel* circuit shown in Fig. 9.1. The lamp is connected in series with a battery of 1.5 V and a resistor of 10 kΩ (much larger than the resistance of the filament). When disconnected from V_s a small current of 150 μA will flow through the filament. As tungsten cools, its resistance decreases. Monitor the time-dependent voltage across the 10 kΩ resistance at a high sampling rate, giving the current $I(t)$, resistance $R(t)$, and temperature of the filament $T(t)$ as a function of time, and providing the possibility of comparison with Newton's law of cooling.

Figure 9.2 shows a plot of the lifetime as a function of applied voltage.

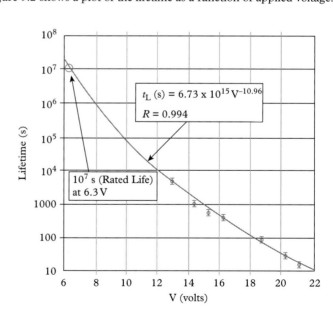

Figure 9.2 *Lifetime as function of voltage (reproduced from Ref [1]).*

Hysteresis effects

Perform I–V measurements when the lamp is operated with a sinusoidal current of frequencies 0.01, 0.1, 1, 10, 100 Hz. Depending on the speed of thermal response of the lamp chosen for study, hysteresis loops of varying widths will appear in the I–V curves. Model the behavior by equating the electrical power to that radiated, conducted, and (if relevant) convected [6].

....

REFERENCES

1. P. Gluck and J. King, Physics of Incandescent Lamp Burnout, *Phys. Teach.* **45**, 466–72 (2008).
2. H. Jones and I. Langmuir, The characteristics of tungsten filaments as functions of temperature, *Gen. Elec. Rev.* **30**, 310, 354, 408 (1927).
3. International Light Technologies, Tungsten halogen lamps – technical/application information, available online at <http://www.intl-lighttech.com/applications/light-sources/tungsten-halogen-lamps>.
4. D. Klipstein, Jr., The great internet light bulb book, available online at <http://members.misty.com/don/bulb1.html#mll>.
5. Wikipedia, Incandescent light bulb, online at <http://en.wikipedia.org/wiki/Incandescent_light_bulb>.
6. D.A. Clauss, R.M. Ralich, and R.D. Ramsier, Hysteresis in a light bulb, *Eur. J. Phys.* **22**, 385 (2001).

10 Propulsion with a solenoid

Introduction 62
Theoretical ideas 62
Goals and possibilities 65
Experimental suggestions 66

Introduction

A solenoid with a ferromagnetic core sliding in and out of it can act as a mechanical actuator in various devices. One can also envisage the use of a solenoid as a kind of electromagnetic gun to propel a ferrous pellet [1]. In both of these applications feasibility and efficiency depend on optimizing the parameters of the pellet and the solenoid for the magnetic force exerted by the latter. This is the subject of the present project.

Theoretical ideas

The field and its gradient

Consider a solenoid of radius a, length l, and number of turns *per unit length n*, through which a current I flows – see Fig. 10.1. The formula for the magnetic field of an infinite solenoid, $B = \mu_0 nI$, is modified for a finite solenoid as follows.

First, use the formula for the field along an axis perpendicular to the plane of a *single* current loop, distance x from its center:

$$B(x) = \mu_0 I a^2 / 2(a^2 + x^2)^{3/2}. \tag{10.1}$$

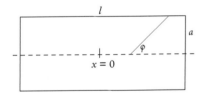

Figure 10.1 *Parameters of the solenoid.*

Now apply (10.1) to a thin section of the solenoid of width dx, with a total current $In dx$, and integrate over the length of the solenoid, using the angle φ rather than x as the variable. Then the on-axis magnetic field is given by [2]

$$B_x = \frac{1}{2}\mu_0 nI (\cos\varphi_2 + \cos\varphi_1), \tag{10.2}$$

where the angles φ_1, φ_2 are shown in Fig. 10.2.

A solenoid's ability to propel a ferromagnetic pellet depends on the inhomogeneity (the gradient) of the field. To see this, consider the pellet to have a dipole moment $\boldsymbol{p}$. The potential energy of a dipole in a field $\boldsymbol{B}$ is $-\boldsymbol{p}\cdot\boldsymbol{B}$, so $\boldsymbol{B}$ acts to align the dipole with the field. The force exerted on a dipole along the axis is the negative of the potential energy's derivative, $\boldsymbol{p}\cdot d\boldsymbol{B}/dx$. Both the field and its gradient inside the solenoid may be measured with a Hall probe or a search coil, and also calculated from (10.2). They will be functions of the aspect ratio, $\alpha = a/l$.

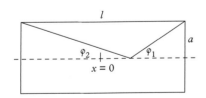

Figure 10.2 *Angle dependence of the field.*

A simple but instructive numerical calculation of **B** and d**B**/dx may be performed by the students themselves in any spreadsheet program, as follows. Take the origin $x = 0$ at the center, as shown in Fig. 10.2. Then *by symmetry* we need only look at **B** and d**B**/dx for half the solenoid. (Symmetry arguments generally deserve our love.) Express the cosines in (10.2) in terms of $\alpha = a/l$ and x/l and calculate numerically both **B** and d**B**/dx from (10.2), for a range of x/l between 0 and 0.7, in steps of say 0.01, and for various values of α. Figures 10.3 and 10.4 illustrate the field **B** and its axial gradient d**B**/dx (in arbitrary units) for several values of α, as functions of x/l from the center to somewhat outside the solenoid ($x/l > 0.5$).

We see that a long narrow solenoid ($\alpha = 0.05$) has a very pronounced field gradient, concentrated near the edge ($x = 0.5$) and dropping to 10% of its peak value within 8% of its length.

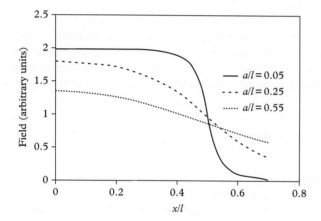

Figure 10.3 *B as a function of x/l for various α.*

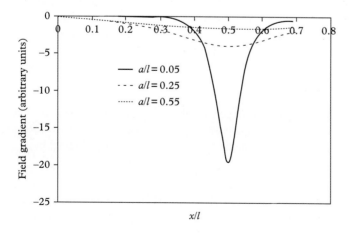

Figure 10.4 *dB/dx as a function of $\alpha = x/l$ for various values of α.*

The energy source

A current pulse of suitable size and duration will ensure that the projectile (in our case, a small ferromagnetic ball or a BB as used in air guns) receives sufficient energy and impulse to take advantage of the gradient of the field while it lasts, enabling the projectile to proceed unhindered through the rest of the coil after it is switched off. One way to do this is to make and break a current source manually, or through a relay switch. Another way is to power the coil by discharging through it a capacitor of capacitance C charged to a potential difference of V. This would enable students to apply the theory of RC circuits and the concept of a time constant. Depending on the relative sizes of the time constants L/R and RC, and of the frequency $1/\sqrt{(LC)}$, the coil inductance L may also play a part. The duration of the pulse will depend on the time constant of the RC circuit, where R is the resistance of the solenoid, since the discharge current is given by $I(t) = I_0 \exp(-t/RC)$. The resistance must be low, so as to provide a large initial current on discharging. This must then be compensated by a large-enough C for the duration of the pulse to be of the order of milliseconds for a typical solenoid length, the time for which the projectile will cover the distance where the field gradient, and therefore the force, is significant. In this setup $d\mathbf{B}/dx$ itself will be a function of time, through the time dependence of the discharge current, $I(t)$. The total effect will be a combination of the two factors. The total energy available for propulsion is $CV^2/2$.

Solenoid design

Since researchers may want to design their own solenoid, the following simple considerations may be helpful. Suppose we have a solenoid of length l and inner radius a. Let it have a multilayer winding of wire having a diameter d, such that the winding thickness is b. Then the number of layers is b/d, and the number of turns per layer is l/d. How much wire do we need? The length of the wire is $2\pi(a + b/2)bl/d^2$. The total number of turns is bl/d^2, so that the number of turns per unit length is $n = b/d^2$. The resistance of the solenoid will be $R = 4\rho l/\pi d^2$, where ρ is the resistivity. If the solenoid is connected to a potential difference of V, the magnetic field will be proportional to $nI = nV/R = d^2V/[8\rho l(a + b/2)]$.

Magnetic force

A soft iron cylindrical rod of permeability μ partially situated inside a long solenoid ($\alpha \ll 1$) will feel a force which may be calculated as follows. Referring to Fig. 10.5, the total self-inductance L is given (neglecting fringing) by

$$L = N^2 A[\mu x + \mu_0(d - x)], \quad (10.3)$$

where the cross-sectional area $A = \pi a^2$. With a current I in the solenoid the magnetic energy is $E = \tfrac{1}{2}LI^2$, and the force $F = dE/dx$ is

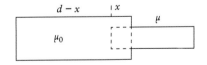

Figure 10.5 *Magnetic force.*

$$F = n^2 I^2 A(\mu - \mu_0)/2, \qquad (10.4)$$

proportional to I^2.

Consider now a spherical soft iron pellet of permeability μ and radius r situated in the region of the large field gradient $d\mathbf{B}/dx$ near the end of the solenoid. The magnetization $\mathbf{M}$, defined as the induced magnetic dipole moment per unit volume, is

$$\mathbf{M} = [(\mu/\mu_0) - 1]\mathbf{B}/\mu_0. \qquad (10.5)$$

The pellet's dipole moment is $\mathbf{P} = V\mathbf{M}$, where $V = 4\pi r^3/3$. In the non-uniform field the force on the dipole is

$$\mathbf{F}_p = \mathbf{P} \cdot d\mathbf{B}/dx. \qquad (10.6)$$

Since $d\mathbf{B}/dx$ varies with x, one may calculate the total impulse acting on the pellet, from some starting position inside the solenoid until it leaves it, by finding $\int F(x,t)\,dt$. This must be maximized for the greatest acceleration and hence exit velocity of the projectile. The result will be a function of the aspect ratio of the solenoid, and will therefore enable one to optimize its parameters, for a given pellet and a given starting firing position.

Goals and possibilities

- Propulsion efficiency (which may be defined as the kinetic energy of the projectile relative to the capacitor stored energy) will depend on the magnetic force exerted on the pellet, which in turn depends on the field gradient. Start by studying the dependence on the current of the magnetic force exerted by the solenoid on an iron core inside. It is also of interest to see how the current required to produce a given force depends on the length of the solenoid. If iron cores of various known permeabilities are available, the effect of permeability on the force may be investigated. Measure the gradient of the magnetic field for solenoids with different aspect ratios.

- The values of length, radius, and number of turns per unit length n (and hence the resulting resistance R and inductance L) of the solenoid should be chosen to maximize the efficiency of the device. The pulsed energy may be supplied from the discharge of capacitors, or from the rapid switching on of a power source. Components should be so chosen that the peak discharge current through the solenoid gives maximum acceleration to the pellet, requiring the conditions for critical damping and operation at the resonance frequency, in the case of a capacitor source.

- In order to achieve the maximum kinetic energy it is important to find the optimum initial position of the pellet near the end of the solenoid, both experimentally and by a model calculation.

- The pellet may be propelled along different kinds of tracks, or launched vertically or horizontally. These will dictate the method of detection employed for measuring the velocity, whether by optical or electromagnetic means, or by simply measuring the distance traveled. Calculate the efficiency of conversion from electromagnetic (capacitor) to mechanical (pellet kinetic) energy.
- Two or more coils in tandem, suitably linked to capacitors and triggered in succession, may be used to increase the velocity of the pellet.

Experimental suggestions

In the first part, several coils of different lengths and number of turns may be prepared, each wound around plexiglass or other nonmagnetic tubing, the cylindrical steel or iron core being somewhat smaller in diameter so it can slide freely in the tube. Possible sizes are 0.03–0.04 m long with 150 turns; 0.065 m long with 280 turns; 0.13 m long with 500 turns; and tube diameter 0.015 m. In order to test the current and voltage dependence of the magnetic force a testing rig may be constructed, as in Fig. 10.6.

Balance the iron core, then attach to the balancing weight some extra weight, and lower the cylinder into the coil with a large current. Gradually reduce the current until the core just slips out of the coil. For this position measure the current I through and the voltage V across the coil. Repeat with a series of additional weights, in order to be able to plot force as a function of current I. Coils of different lengths may tested in order to see how the *efficiency*, defined as the ratio of the induced magnetic field to the supplied current, varies as the ratio of length to radius increases.

One may explore the field gradient as follows. Fix the current. Hang the cylinder, situated in the center of the solenoid, on a force sensor. Then move the solenoid gradually away from the cylinder and measure the force as a function of position.

For propulsion, one possibility is to use a glass tube as the track. (A length of 1.2–1.5 m is suitable; its diameter will depend on the pellet used. Using tubes of, say, 0.02 m in diameter will trap a large flux into a small area.) Two or more capacitors of 4500 μF rated at 50 V with suitable resistances connected to a variable power source may serve as the energy source. Use a suitable (knife) switch, or auto starter relay. If shooting horizontally, place onto wooden support brackets. To measure the exit speed of the pellet, use two short coils wound around the tube near each other and connected to an oscilloscope or a timer. Alternatively, see how far the pellet travels before hitting a sheet of carbon paper on the ground below. Or, the pellet may be fired through a vertically held tube and the height reached may be measured by inserting a loosely fitting paper in the tube to see how far it is propelled. Another option would be to use a photogate positioned just beyond the end of the solenoid.

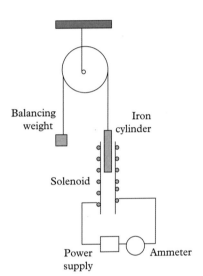

Figure 10.6 *Testing rig.*

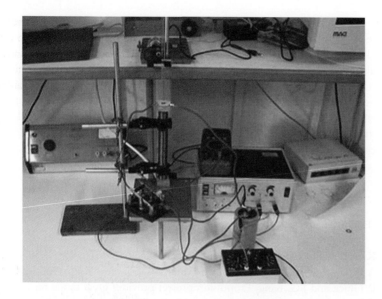

Figure 10.7 *Vertical magnetic propulsion apparatus.*

When optimizing the pellet position, use a graduated non-magnetic plunger in order to measure its initial position from the end of the solenoid. A non-magnetic screw would make adjusting the initial position easy, knowing the distance traveled as it advances by one turn. Plot the resulting velocity as a function of this position.

Figure 10.7 shows a vertically mounted 20 cm-long solenoid, 900 turns in 6 layers, with resistance 1.1 Ω. It surrounds a 0.5 cm-diameter 20 cm-long glass tube [3]. At the solenoid exit there is a somewhat wider tube through which the pellet continues upward after being launched. At the bottom of the wider tube is a photogate which measures the velocity of the pellet. The pellet here was 4.3 mm in diameter. It rested in a small depression on the top of a vertically movable non-magnetic screw, as shown in Fig. 10.8. The pellet was propelled by discharging a 4000 μF capacitor charged to 150 V.

Figure 10.9 shows the velocity of a pellet as a function of its position from the bottom end of the solenoid [3]. As the position increases, the magnetic field gradient decreases, and so the force, the acceleration, and hence the exit velocity will also decrease.

The force on the pellet at any instant is $\boldsymbol{p} \cdot d\boldsymbol{B}/dx$, but the magnetic field gradient is itself a function of the capacitor discharge current $I(t)$. Since the pellet leaves the area where the field gradient is significant within a millisecond or so, the dominant influence is $d\boldsymbol{B}/dx$ rather than $d\boldsymbol{B}/dt$. A particularly gung-ho researcher might wish to think about constructing a numerical simulation, as an *analytic* solution for the exit velocity would be complicated and requires determining the instantaneous force on the pellet (which is a function of the fast decaying current), and its cumulative effect as it passes through the solenoid after the launch. This would also require knowledge of the magnetic permeability of the pellet, the

Figure 10.8 *Movable pellet.*

magnetic field inside it, and a calculation of its dipole moment [4,5]. In the absence of such information, the magnetic permeability could simply be taken as an adjustable parameter. Here we merely outline the procedure: a full numerical simulation would take into account the cumulative effect of $I(t)$, at least until the pellet has moved 8–10% of the solenoid length. From the instantaneous force one may calculate the instantaneous acceleration at any x. Its cumulative effect will give the exit velocity of the projectile. The exit velocity of the projectile will be a function of the aspect ratio of the solenoid, the current in it, and the duration of the energizing pulse via the time constant.

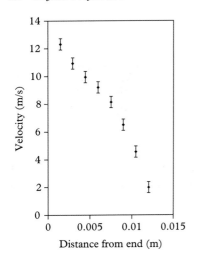

Figure 10.9 *Exit velocity as a function of distance from the end.*

REFERENCES

1. K.P. Williams, Electromagnetic coil launcher, *Nuts and Volts* **29**(3), 50, 2008.
2. A.F. Kip, *Electricity and Magnetism* (McGraw-Hill, New York, 1962), p. 119.
3. P. Gluck and J.G. King, Propulsion project provides great scope for investigating electromagnetism, *Phys. Educ.* **45**, 466 (2008).
4. I.S. Grant and W.R. Phillips, *Electromagnetism* (Wiley, New York, 1982).
5. R. Fitzpatrick, A uniformly magnetized sphere, 2002, available online at <http://farside.ph.utexas.edu/teaching/jk1/lectures/node50.html>.

Magnetic dipoles

11

Introduction

Although microscopic magnetic dipole moments play an important role in atomic spectra, NMR, medical imaging, and solid state physics, experiments with them cannot be performed with simple equipment, or appreciated at the level of introductory courses in electromagnetism. But since the latter always discuss the concept, at least for a small current-carrying loop, the project suggested below, which involves macroscopic dipoles and their interaction with magnetic fields, can aid in understanding many of the ideas introduced in electromagnetism.

Introduction	69
Theoretical ideas	69
Goals and possibilities	71
Experimental suggestions	72

Theoretical ideas

A bar magnet of length L may be thought of as consisting of two magnetic poles or "charges" $q_m^\pm$ of opposite sign situated at or near its two ends, and so separated by nearly L. The effective position of these poles may be determined by experiment using iron filings. The bar magnet may be considered to be an approximate magnetic dipole, of moment $p = q_m L$. The finite size of the bar magnet needs to be taken into account when performing experiments on the interaction of such a magnet with external fields, or the interaction of two such magnets in which the relevant distance scales are comparable to L.

In magnetostatics, the radial magnetic field generated by $q_m^\pm$ at a distance r is [1,2]

$$B_q^\pm = \mu_0 q_m^\pm / 4\pi r^2. \tag{11.1}$$

The potential energy of a "point" dipole of moment p in an external magnetic field B which tends to align p along it is $U = -p \cdot B$. The torque exerted on p is $p \times B$.

The magnitude of B along an axis perpendicular to the plane of a circular coil of radius R, having N turns, and carrying a current I, and going through its center, at a distance x from that plane, is give by, as in (10.1),

$$B_{\text{coil}} = \mu_0 I N R^2 / 2 \left(x^2 + R^2\right)^{3/2}. \tag{11.2}$$

Let a very short bar magnet of moment p and mass m be placed on the coil axis. It will execute oscillations along the axis about the coil center. Assuming

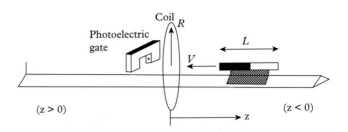

Figure 11.1 *Magnet passing a coil.*

that these are small, namely if $x/R \ll 1$, its potential energy may be expanded in powers of x/R to give

$$U(x) = -\mu_0 I N p / 2R + [3\mu_0 i N p / 4R] x^2 / R^2 - \cdots, \qquad (11.3)$$

where the first term is the energy at the center of the coil. The frequency of oscillation is calculated from the second, harmonic term, and higher-order anharmonic terms can be neglected for $x/R \ll 1$. The frequency is given by

$$f^2 = 3\mu_0 I N p / (8\pi^2 m R^3). \qquad (11.4)$$

This is only an approximate formula and must be corrected by allowing for the finite length L of the bar magnet, since that is comparable to the amplitude of oscillations. We shall only outline the method with which one calculates the potential energy, not of a point dipole but one of length L. If l is a length along it, then the amount of dipole for a length dl is $dp = p \, dl/L$. Let b be the distance from the coil center to the bar center. Then

$$dU = -B_{b+1} \, dp, \text{ and } U = \int_{-L/2}^{+L/2} dU. \qquad (11.5)$$

The rest of the calculation proceeds as before by a power series expansion of U.

Suppose now that the coil does *not* carry a current and that the bar magnet of pole strength $q_m^\pm$ approaches the coil with constant velocity V on a frictionless air track, as shown in Fig. 11.1 [3]. The induction pulse thus generated is calculated from the Faraday law $\varepsilon = -d\Phi/dt$, where Φ is the flux of the magnetic fields in (11.1) through the coil. It may be calculated by using the convenient imaginary hemispherical surface shown in Fig. 11.2. One obtains, for the induced time-dependent EMF pulse [3]:

$$\varepsilon = \frac{V N \mu_0 R^2 p}{2L} \left\{ 1 \Big/ \left[(V(t-t_0) + L/2)^2 + R^2 \right]^{3/2} - 1 \Big/ \left[(V(t-t_0) - L/2)^2 + R^2 \right]^{3/2} \right\}, \qquad (11.6)$$

where $z = V(t - t_0)$, and at t_0 the center of the magnet is exactly at the coil center, where $\varepsilon = 0$.

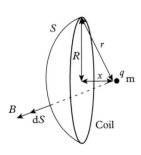

Figure 11.2 *Flux calculation.*

The area under the $\varepsilon(t)$ curve, namely,

$$\int_{t_0}^{\infty} \varepsilon(t)\,dt, \tag{11.7}$$

is a constant independent of the glider-magnet velocity, as may be shown by integration of (11.6). Thus, higher velocity means increasing pulse height and decreasing pulse width.

Goals and possibilities

- Mount a small bar magnet on top of a glider on an air track. Let a circular current-carrying coil encircle the track with its plane perpendicular to the track and its center at the height of the long axis of the magnet. Study the dynamics of the magnet as it executes small oscillations about the center of the coil. Calculate the frequency of these oscillations, taking into account the finite extent of the magnet and its interaction with the magnetic field of the coil.

- Measure the magnetic moment of a short magnet by timing its oscillations in a constant (Helmholtz coil) magnetic field. The magnet could be a compass or dip needle, or a short bar magnet of circular or rectangular cross section for ease of calculating its moment of inertia. The magnetic moment of the magnet on the glider may also be found by using the formula for its own axial magnetic field,

$$B_p = \frac{\mu_0 m}{4\pi L}\left\{\frac{1}{(x-L/2)^2} - \frac{1}{(x+L/2)^2}\right\}, \tag{11.8}$$

measuring its field B_{exp} for a series of values of x experimentally, large enough compared to the value of L, by a Hall probe, and comparing B_p with B_{exp}.

- Measure the magnetostatic force law $F = k/r^n$ between two short cylindrical bar magnets, where k is some constant for the two magnets and r denotes some typical distance between them. Is this r the distance between the centers of the magnets, or is it somehow to be modified to take into account both magnetic repulsion and attraction bteween the poles?

- Let the circular coil not carry a current but be connected to a sensitive storage oscilloscope or a high speed data acquisition system. Launch the magnet dipole with the glider at various reproducible speeds at and through the coil. Record the rise and fall of the EMF induced in the coil, the induction pulse, as a function of the launch velocity, as the magnet approaches and recedes from the coil. Given the geometrical characteristics of the coil and the moment of the magnet, calculate the shape of these time-dependent

EMFs and compare with experiment. Integrate the experimental induction pulse curve to show that the area under half the pulse is a constant independent of velocity.

- One could also study propulsion of magnets on an air track by current-carrying coils, the air track serving as an alternative to the apparatus described earlier in the magnetic propulsion project of this chapter.
- Dipole interactions on an inclined airtrack could also be a project extension – see [4].

Experimental suggestions

When measuring the dipole moment of the magnet from the period of its oscillation in a magnetic field, make sure that the bar magnet is small compared to the diameter of the Helmholtz coils, in order that its oscillations take place in the nearly uniform magnetic field at the center of the coils. If necessary, determine the torsional modulus of the thread supporting the magnet.

In order to determine the force law between two identical cylindrical bar magnets, place them with identical poles facing each other in a glass or plexiglass tube whose diameter is slightly larger than that of the magnets, as in Fig. 11.3. A suitably arranged digital balance, or aluminum rods of known weight placed on the top magnet, may be used to exert known forces and thus to vary the distance between them. In analyzing the data you should make allowance for the fact that the magnetic poles are at some distance h from the ends, so that the relevant separation is actually $r + 2h$. In addition, you must take into account both the mutual repulsion and attraction of the near and far poles of the magnets.

A rather appealing alternative method would be the following: let the top magnet float above the bottom one at some equilibrium separation. A function generator connected to a solenoid wound around the tube will produce fields that exert forces causing small oscillations of the top magnet around its equilibrium position. At the resonance frequency the amplitude will be largest and this can be readily observed. This frequency is related to the net restoring force acting on the magnet, a result of its weight and the mutual repulsive force between the magnets, from which the force law may be deduced.

For the experiments on the air track, a coil of radius 20 cm and about 150 turns is suitable. In the oscillating magnet experiment the coil should be connected to a stable current supply. Exact alignment of the axis of the magnet mounted on the glider with the axis of the coil is important. Photogates may be used to measure the oscillation frequency as a function of the current through the coil.

In the induction experiment a pair of photogates may be used to measure the velocity of the glider with the magnet mounted on it. The width of the photogate gate beam must also be taken into account for accurate measurements. There are launching mechanisms of the air track gliders that give reproducible launch velocities, but this should be checked. If a combination of the theoretical magnetic

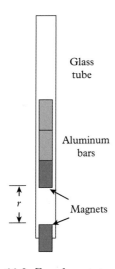

Figure 11.3 *Force law setup.*

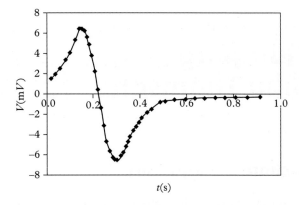

Figure 11.4 *Induction pulse.*

field of the dipole and Hall probe measurements of its field is used to deduce the dipole moment of the magnet, make sure that the Hall probe measures the field at points not too close to the magnet. The induction pulse in the coil may be recorded by a computer data acquisition system sampling at millisecond rates. The time t_0 in (11.6) is determined from the condition $\varepsilon(t_0) = 0$ and can be obtained from the linear portion around $\varepsilon = 0$ of the experimental $\varepsilon(t)$ curve, shown in Fig. 11.4 (after [3]).

....

REFERENCES

1. D.E. Roller and R. Blum, *Physics* (Holden Day, San Francisco, 1981), vol. 2.
2. Experiment 4, Forces and torques on magnetic dipoles, *Electricity and magnetism*, MIT course 8.02, online at <https://www.edx.org/course/mitx/mitx-8-02x-electricity-magnetism-608>.
3. J.A. Manzanares, An experiment on magnetic induction pulses, *Am. J. Phys.* **62**, 702 (1994).
4. H.G. Lukefahr, Magnetic dipole interactions on an air track, *Am. J. Phys.* **60**, 1134 (1992).

12

The jumping ring of Elihu Thomson

- Introduction 74
- Theoretical ideas 74
- Goals and possibilities 75
- Experimental suggestions 76

Introduction

The Elihu Thomson jumping ring apparatus is often used as a spectacular, though only qualitative, demonstration of AC theory and electromagnetic induction. In fact, one may explore quantitatively many of the parameters which govern the behavior of the ring. The physics of induction, levitation, motor, transformer, three-dimensional magnetic field patterns, shaded pole motor, and RL and RLC circuits may all be relevant, and contribute to making this project rich in both theory and measurements. It is recommended that the researcher design and build the equipment, where possible.

Theoretical ideas

See Refs [1–4]. The apparatus consists of a solenoid containing a long iron core, and an aluminum or copper ring. The ring has both resistance R and inductance L. The latter is responsible for a phase lag between the current induced in it and the solenoid EMF, so creating an average force between ring and iron core that is *repulsive*. The force is proportional to the square of the solenoid current and the square of the frequency, the number of turns per unit length, and inversely proportional to the ring's impedance, $\sqrt{(\omega^2 L^2 + R^2)}$. Geometrical factors such as the radius, thickness, and width of the ring affect its R and L.

Let I_s and I_r be the currents in the solenoid and the ring, respectively. The current in the solenoid is AC:

$$I_s = I_0 \sin \omega t. \tag{12.1}$$

The magnetic field B_s in the solenoid will have the same time dependence. The mutual inductance M of the ring and the solenoid (near the) center is

$$M = \mu_0 n N \pi a^2, \tag{12.2}$$

where n, N, and a are the number of turns per unit length in the solenoid, the number of turns (= 1), and the radius of the ring, respectively. The ring and the solenoid together may be thought of as a simple transformer, with the solenoid playing the role of the primary and the ring acting as the secondary. The radial component of B_s is $B_\rho = \mu_0 n I_s K$, where K is a numerical factor which takes into account the specific materials and geometry and their effect upon the fringing of the solenoid field. The force of repulsion on the ring is thus

$$F = 2\pi a I_r B_\rho. \tag{12.3}$$

By Faraday's law, the axial component of the oscillating magnetic field passing through the ring induces in it a current. The ring can be modeled as a simple LR circuit of the sort treated in introductory physics textbooks. One can show that the induced current equals

$$I_r = \left[MI_0\omega / (R^2 + \omega^2 L^2)\right] (R\cos\omega t + \omega L \sin\omega t). \tag{12.4}$$

Combining (12.2), (12.3), and (12.4) we get

$$F = (2\pi a\mu_0 nK) \left[M\omega I_0^2 / (R^2 + \omega^2 L^2)\right] (R\sin\omega t\cos\omega t + \omega L\sin^2\omega t).$$

If the ring is in a stable position, and if we average the force over a complete period, we obtain

$$\langle F \rangle = (\mu_0 M K n\pi a)\omega^2 \, LI_0^2 / (R^2 + \omega^2 L^2). \tag{12.5}$$

The force is proportional to the frequency squared. The inductance L of the ring plays a decisive role in producing the force. Since both K and M depend on height, the force is not a constant. At the center of a symmetric solenoid $K = 0$, so that KM is a minimum. As one goes up from the center KM rises to a maximum in the vicinity of the top of the solenoid, then decreases again. In the limit of high frequencies, where $I_r R \ll L dI_r/dt$, the force is independent of the frequency:

$$\langle F \rangle = (\mu_0 n K M \pi a / L) I_0^2. \tag{12.6}$$

If the ring alone is replaced by a circuit containing, in addition, a capacitor C, the factor ωL is replaced by $\omega L - (\omega C)^{-1}$, and provided the size of C is right, the repulsion on the ring could be replaced by an attractive force. The required size of C may be affected by the resistance of the leads connecting it to the circuit.

Goals and possibilities

- Measure the (average) force on the ring as a function of the alternating current's amplitude and frequency.
- Measure the (average) force as a function of the distance between the ring and the solenoid.

- Measure the oscillating force on the ring with a force transducer, as well as its phase relative to the current.
- Map both the vertical and horizontal components of the magnetic field of the solenoid. Relate the findings to the force data.
- Measure the height of the jump as a function of the voltage.
- Study the variation of the height of levitation with the applied voltage.
- Explore the effects of ring geometry and material on all of the foregoing.
- Alter the usual apparatus by incorporating a variable capacitance C into the ring's circuit, whereby the reactance of C will influence markedly the motion of the ring, and may even reverse its direction.

Experimental suggestions

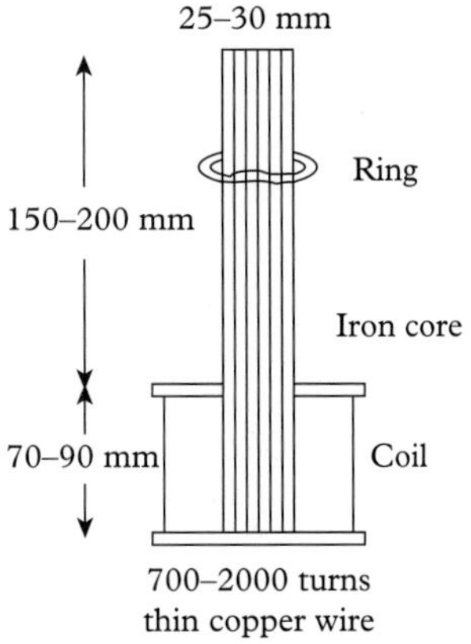

For the solenoid an ohmic resistance of 40–60 Ω is useful, using some 700–2000 turns of enamelled SWG 20–24 wire, on a plastic or wooden former 7–10 cm high, with a cylindrical hollow central shaft of diameter 25–40 mm. The latter accommodates a tightly fitting 20–30 cm-long plastic pipe which is filled with an iron core along its length. The core can be made from iron wires. Refer to Fig. 12.1. Aluminum and copper rings of varying width, thickness, and radius can be cut from suitable pipes.

The AC source can be an adjustable auto-transformer, delivering 5–8 amperes and connected to the mains. To investigate the frequency dependence of the force on the ring, a signal generator's output may be amplified with the heavy duty hi-fi audio amplifier of a public address system with a suitable line output, and connected to the solenoid. A heavy duty toggle switch should be incorporated into the circuit, enabling one to induce sudden current increases to make the ring jump.

Figure 12.2 shows a convenient arrangement for static force measurements. The solenoid and core are supported upside down. An aluminum or copper ring is fastened to the top of a suitable plastic or cardboard tube, which rests on an electronic scale. The scale gives fast readings, so that large currents need not be maintained beyond a few seconds. The tube is long enough so the scales are not influenced by the magnetic field. The scales may be placed on a jack, enabling one to vary the ring–solenoid distance.

If, instead of resting on scales, the bottom of the tube supporting the ring is carefully attached to a force probe from a data acquisition set, one can measure the oscillating force as a function of time via the computer interface, and compare its phase and frequency with that of the solenoid current.

With the apparatus in its normal position the height of the ring's jump (on sudden switching) can be measured as a function of the voltage applied; the dependence is roughly linear. The ring will levitate on the tube if one increases the voltage slowly from zero to the same value as before. The height of levitation

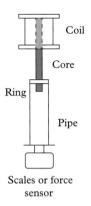

Figure 12.1 *Solenoid and iron core.*

Figure 12.2 *Setup for static force.*

as a function of voltage may be investigated, and turns out to be approximately quadratic. As part of the project, it is clearly of interest to predict these behaviors with a theoretical model.

A Hall probe may be used to map the magnetic field around the solenoid. Alternatively, one may slip a coil of a few turns on the core so that it can slide along it, as in Fig. 12.3, and measure the induced voltage in it (reflecting the vertical component of the field) with a multimeter or oscilloscope. Touching the core with the coil when the latter's axis is horizontal enables one to map the horizontal component of the magnetic field.

If the ring is made of kitchen aluminum foil, no jumping occurs: the resistance is too large. But if a thick aluminum ring is also placed over top of the core, the foil jumps up to meet it: a result of the attractive force between the two produced by induction.

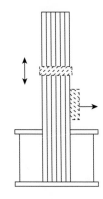

Figure 12.3 *Field mapping.*

With the solenoid and core in the upside-down position, the shaded pole motor principle can be explored. A copper or aluminum disk, suspended so that it can rotate freely about its axis, is held just above the solenoid – see Fig. 12.4. If a metal sheet protrudes partially over the solenoid, the disk will rotate, due to the attraction resulting from the current induced in the sheet and the unshaded portion of the disk. Without the shading sheet there is no rotation, only a repulsion of the disk.

Connecting a capacitance to the ring results in it being attracted to the solenoid, as mentioned in [1]. Detailed suggestions for carrying out this part of the project may be found in [4]: the aluminum ring is replaced by a small coil of 40 loops with a resistance R of $0.06\,\Omega$ and inductance $L = 0.65\,\text{mH}$ ($R \ll \omega L$). With the solenoid and core horizontal, the small coil is slipped on the core, and is suspended by (non-conducting) fine threads, as shown in Fig. 12.5. Various capacitances C (thousands of microfarads, e.g. a frequency of 200 Hz for $C = 4000\,\mu\text{F}$) can be connected to the coil by lightweight wires (so as not to disturb the coil's motion). Depending on C and ω the secondary circuit can be inductive, capacitative, or merely resistive. At resonance, when $\omega L = 1/\omega C$, the coil does not move when the AC current is switched on in the solenoid. As C is reduced ($1/\omega C$ is somewhat larger than ωL, so that the circuit becomes capacitative), the coil swings towards the solenoid, rather than away from it. The force is proportional to the product of the primary and secondary currents, and becomes attractive rather than repulsive, because of the phase relationship between the primary and secondary currents under capacitative conditions.

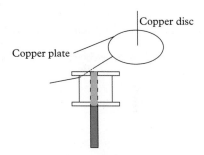

Figure 12.4 *Shaded pole motor.*

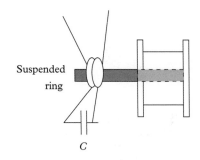

Figure 12.5 *Capacitor causes attraction.*

The graphs in Fig. 12.6, reproduced from a typical experiment in [5], show the dependence of the induced secondary voltage on the AC voltage applied to the core, for both the horizontal and vertical positions of the plane of the coil.

Figure 12.7 shows the dependence on the AC voltage of both the height jumped by the ring and the height of levitation [5].

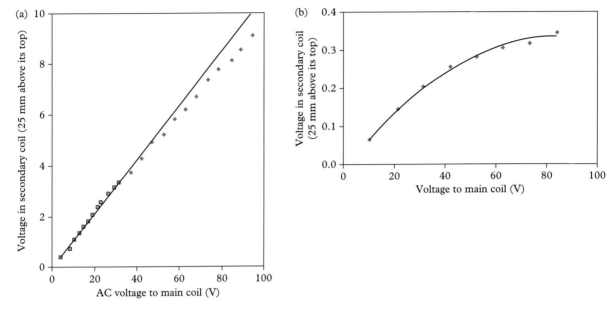

Figure 12.6 *Secondary voltage as a function of main coil voltage: (a) vertical component; (b) horizontal component.*

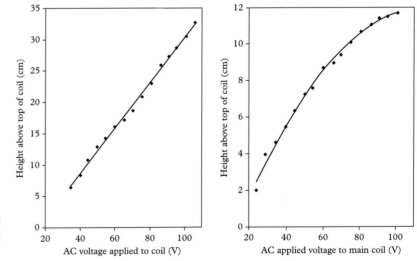

Figure 12.7 *Height of jump (left) and of levitation (right) as function of applied voltage.*

REFERENCES

1. W.M. Saslow, Electromechanical implications of Faraday's law, *Am. J. Phys.* **55**, 986–93 (1987).
2. S.Y. Mak and K. Young, Floating metal ring in an alternating magnetic field, *Am. J. Phys.* **54**, 808–11 (1986).
3. A.R. Quinton, The ac repulsion demonstration of Elihu Thomson, *Phys. Teach.* **17**, 40–42 (1979).
4. B.S. Perkalskis and J.R. Freeman, Extending Elihu Thomson's demonstration and Lenz's law, *Am. J. Phys.* **65**, 1022–4 (1997).
5. D.J. Sumner and A.K. Thakkrar, Experiments with a jumping ring apparatus, *Phys. Ed.* **7**, 238 (1972).

13 Microwaves in dielectrics I

Introduction 80
Theoretical ideas 80
Goals and possibilities 81
Experimental suggestions 82
Further explorations 85

Introduction

The absorption and transmission of electromagnetic radiation in materials is a broad subject. Losses may be high or low; dielectric constants can vary over a large range of values and depend on frequency; wavelengths will change on traversing the material; exponential attenuation may occur; or standing wave patterns may be created. It is interesting to compare and contrast the behavior of light and microwaves for the same material. The passage of microwaves through materials characterized by different loss mechanisms may also be explored. The real and imaginary parts of dielectric constants may be measured, and the amount of attenuation in the incident radiation could be investigated for non-invasive testing of some properties of the medium. The present chapter, and the following one, are devoted to some of these topics.

Theoretical ideas

All the experiments we suggest will have radiation incident perpendicular to slabs of material with parallel faces of thickness d. Let n, k, λ be the real and imaginary parts of the refractive index and the wavelength of the radiation, respectively, where n and k are related to the relative permittivity and conductivity. The fraction of radiation transmitted T is given by [1–3]

$$T = \frac{A}{B^2[E + (C/B)^2/E - 2C\cos(\delta)/B]}, \quad (13.1)$$

where

$$A = 16(n^2 + k^2),$$
$$B = (n+1)^2 + k^2,$$
$$C = (n-1)^2 + k^2,$$
$$E = \exp(4\pi kd/\lambda),$$
$$\delta = 2\theta + 4\pi nd/\lambda,$$
$$\tan\theta = \frac{2k}{n^2 + k^2 - 1}.$$

If there is little or no loss (k being small) then thin film interference patterns may be established as the thickness of the material d is increased. These may be observed by increasing d in steps that are small compared to the wavelength, which is around 3 cm for the microwave equipment normally available in teaching laboratories. Varying the thickness may be achieved by stacking identical plates whose thickness compared to the wavelength is judiciously chosen. The pattern will show up as a periodic variation of $T(d)$ as a function of thickness. The expression (13.1) for T depends on d and has two adjustable parameters n and k. These may be varied in order to obtain the best fit to the experimental measurement of $T(d)$. The numbers thus obtained will then be the experimental values for n and k, which may be compared to those available in handbooks [4,5] for the materials chosen.

In the microwave region, handbooks provide values of the relative permittivity ε_r and the electrical conductivity σ, which are related to n and k, at a frequency ω, by the relations [2,3]:

$$2n^2 = \varepsilon_r\left(1 + \sigma^2/\omega^2\varepsilon_r^2\varepsilon_0^2\right)^{1/2} + \varepsilon_r, \quad 2k^2 = \varepsilon_r\left(1 + \sigma^2/\omega^2\varepsilon_r^2\varepsilon_0^2\right)^{1/2} - \varepsilon_r. \qquad (13.2)$$

For materials in which k is large (water, electrolytes, wood) the transmission coefficient will show an exponential attenuation with thickness: both the wavelength (on account of $n > 1$) and the wave amplitude (on account of k) decrease inside the medium. The periodic pattern may then be difficult to detect.

By contrast, since for light the wavelength is five orders of magnitude smaller, even a perfectly transparent dielectric will not show any periodic behavior for $T(d)$.

Goals and possibilities

- Measure the dielectric constant of castor oil by determining the change in wavelength of a standing microwave pattern in it as compared to that in air, using a reflecting surface between transmitter and reflector.
- Measure the coefficient of transmission of normally incident microwave radiation through an increasing number of plates of glass, lucite, and other materials, stacked closely together. Use the theory in order to extract for these materials experimental values of the real and imaginary parts of the refractive index.
- Distinguish between low- and high-k materials, and between periodic (standing waves) and exponentially decaying behavior of $T(d)$.
- Distinguish between light and microwave behavior by passing light through stacks of microscope slides and measuring quantitatively the ratio of transmitted to incident light intensity.

Experimental suggestions

It is assumed that the laboratory has at its disposal a microwave set used for demonstrating wave properties. Do not use a microwave oven, it is not safe.

Refractive index of castor oil

See Ref. [6]. Use a thin-walled plastic container $30 \times 15 \times 15$ cm to hold the liquid. A possible standing wave setup is shown in elevation in Fig. 13.1.

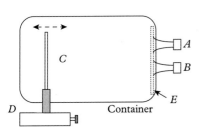

Figure 13.1 *Setup for standing waves in oil.*

A, B, C, D, and E are, respectively, the microwave transmitter, the receiver, a 12×7 cm metal plate completely immersed in the liquid in order to reflect the microwaves, a travelling microscope carriage to move C, and a metal screen rising above the liquid surface. The important role of E is to shield C's support from incident radiation, so as to prevent spurious reflections from it. Make sure that C can be moved to and fro in a controlled manner 3–4 wavelengths in the liquid, and that 2–3 wavelengths separate it from the container's sides. Moving C, plot the intensity of the receiver as a function of C's distance from some reference point, both in air and when oil is in the container. The dielectric constant can be extracted from the ratio of wavelengths and from the relation $\lambda f = 1/\sqrt{(\mu\varepsilon)}$, where μ is the permittivity, which to a good approximation may taken to be that for air, $4\pi \times 10^{-7}$ H/m.

Figure 13.2 shows typical data, reproduced from [6], for standing wave patterns for castor oil and air. The horizontal axis is the position of the reflecting surface from some arbitrary reference point; the vertical axis is the detected amplitude in arbitrary units.

Microwave transmission in lossless media

See Ref. [7]. Procure 20 identical glass plates, large enough to rest on the top of the detector's horn, and of identical thickness (between 1 and 1.5 mm), and a

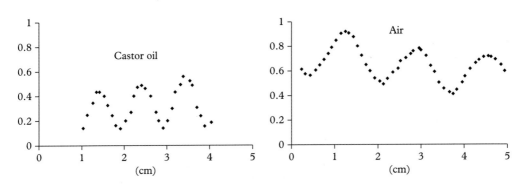

Figure 13.2 *Standing waves: amplitude versus distance.*

similar set of plates made of lucite (perspex). It may be useful to have some means of varying the emitted microwave intensity, say by placing a rotatable metal grid of parallel wires in front of the transmitter horn. A simple setup is suggested in Fig. 13.3. Tilt the set of plates or the transmitter from the vertical so as not to return the waves to the transmitter.

The horn-to-horn distance should be large enough to justify the assumption of plane waves reaching the detector. Meter readings are recorded as each dielectric plate is placed on the detector. The transmission data may then be analyzed in conjunction with the theoretical formula in order to fit it best with suitable values of n and k.

Figure 13.4 shows the percentage transmission as a function of the number of glass plates [2,3].

Microwave transmission in an electrolyte

See Ref. [8]. The setup from the previous experiment may be used, except that the transmitter is placed *under* a thin-walled container (say a small aquarium) holding an amount of salt solution, of carefully measured concentration, that is just enough to cover its bottom. This being the reference level, the meter reading is recorded. Carefully measured quantities can then be injected by a syringe in order to increase the thickness of the electrolyte by a millimeter each time (it helps to know the surface area of the container), and measurements of the transmitted power can be recorded (*assuming* a square-law detector).

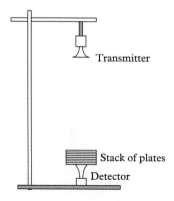

Figure 13.3 *Transmission.*

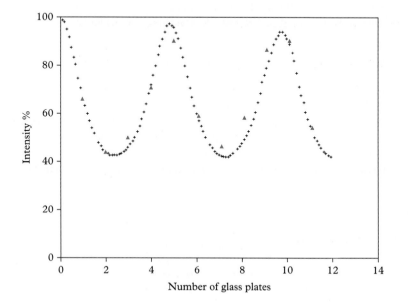

Figure 13.4 *Transmission as function of number of plates: theoretical and experimental points.*

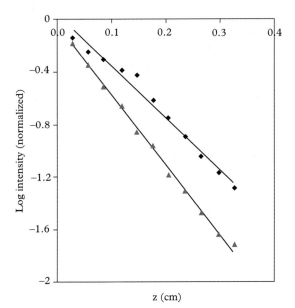

Figure 13.5 *Intensity as a function of thickness for water (squares) and salt solution (triangles).*

The detector should be shielded with some care from unwanted reflections. This may be achieved by suitable placement of metal sheets around it. One may start the investigation with tap water, then continue with increased salt concentrations (which increases the conductivity), in order to see whether at microwave frequencies this affects the attenuation.

Figure 13.5 shows the results of a typical experiment [8] for the intensity as a function of thickness z reckoned from the top surface of the liquids.

Transmission of light through stacked plates

A parallel beam of light from a strong source, directed through an aperture, passes through a series of plates and is then detected by a photodetector whose output is connected to a suitable recorder (an oscilloscope or a data acquisition sysyem), as shown in Fig. 13.6. Sets containing 1–15 identical plates may be obtained, say 5 × 5 cm, of various materials (glass, perspex, clear acetate sheets from an overhead projector, microscope slides) and widths (from a fraction of a millimeter

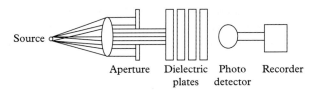

Figure 13.6 *Light transmission setup.*

to 30 mm), and tested for transmission intensities, either singly, for the thickest, or a whole series in tandem, for the thinnest. The photodetector should be calibrated [7] for a linear response for the intensities of interest. If various widths of the same material are available, one can attempt to quantify the absorption through them as a function of thickness according to an exponential law. For very thin sheets this effect is not important and then (13.2) may serve as the basis for comparison with theory.

Further explorations

Readers wishing to delve more deeply into transmission and reflection of microwaves will benefit from consulting Ref. [9–11]. Reference [9] could serve as inspiration for extending the project, or as an independent project.

REFERENCES

1. G. Jungk, Determination of optical constants: The plane parallel slab, *Appl. Opt.* **19**, 508 (1980).
2. F. Thomson and H. Tsui, Transmission of normally incident microwave radiation through parallel plates of material, *Am. J. Phys.* **54**, 712 (1986).
3. R. Coelho, *Physics of Dielectrics* (Elsevier, Amsterdam, 1979), p. 61.
4. T. Moreno, *Microwave Transmission Design Data* (Dover, New York, 1948), p. 408.
5. D.M. Pozar, *Microwave Engineering* (Wiley, New York 2005).
6. W.P. Lonc, Microwave propagation in dielectric fluids, *Am. J. Phys.* **48**, 648 (1980).
7. E.F. Zalewski, NBS measurement services: The NBS photodetector spectral response calibration transfer program, available online at <http://www.nist.gov/calibrations/upload/sp250-17.pdf>.
8. Y.H. Haranas and W.P. Lonc, Microwave attenuation in liquid media, *Am. J. Phys.* **52**, 214 (1984).
9. R. Rulf, Transmission of microwaves through layered dielectrics – theory, experiment and applications, *Am. J. Phys.* **56**, 76 (1988).
10. W.L. Schaich, G.E. Ewing, and R.L. Karlinsey, Influence of multiple reflections on transmission through a stack of plates, *Appl. Opt.* **45**, 7012 (2006), available online at <http://www.ncbi.nlm.nih.gov/pubmed/16946779>.
11. R. Baets, *Photonics*, course syllabus, ch. 3, Universiteit Ghent, available online at <http://www.photonics.intec.ugent.be/download/ocs127.pdf>.

14 Microwaves in dielectrics II

Introduction

We know that the basis of microwave cooking is the large absorption of microwave energy by water molecules in food, with this energy being subsequently transferred to the rest of the ingredients. One can therefore use microwave absorption to measure the moisture content of materials quantitatively; since this has relevance to soil mechanics, part of this applied project is to measure the water contents of various soils and clays [1–4].

Absorption and transmission intensities depend both on material density and water content, as well as on inhomogeneities present in the material. In the second part of the project this dependence is used to investigate the dryness and density of various timbers, and to reveal hidden holes in slabs of wood. The idea is that if the hole sizes are of the same order as the radiation wavelength, then microwaves passing through such timber would be both absorbed and *diffracted* by the holes, affecting the transmitted angular distribution of the radiation.

Experimental suggestions

Water content of soils

The experimental setup is shown in Fig. 14.1. Prepare a sample holder from $1/4''$ thick plastic, the space for the specimen some $4 \times 1/2''$ thick. Uniform grain sand is dried in an oven until its weight is constant, when it is moisture free. It is inserted closely packed into the sample holder, and microwaves are then passed through it. The intensity reading on the detector will be the reference level for dry sand. One can then prepare samples with increasing water content, thoroughly mixed, packed tightly into the holder, and measure the transmitted intensity as a function of water content. About 12% water by weight is the upper limit, beyond that water starts separating out and the sample ceases to be homogeneous. Figure 14.2 shows the results of the experiment reported in [1], where the transmitted intensity is plotted as a function of the water content of the sand, with an average grain size of 0.3 mm.

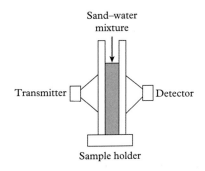

Figure 14.1 *Moisture content setup.*

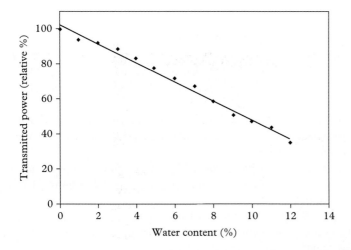

Figure 14.2 *Intensity as a function of water content of sand.*

Absorption and defects in timber

Moisture, density, and thickness of wood are parameters that can be measured by microwave transmission experiments. In addition, since wood is anisotropic, it is interesting to see the effect on absorption of wood fibres parallel or perpendicular to the incident microwave beam. Absorption as a function of thickness may be investigated by placing a series of 1 cm-thick slabs in the path of the beam. A more challenging task is to bore holes of different measured diameters a and lengths b in blocks of wood made up of joined panels, to find out how these parameters affect the transmitted microwave intensity, and then to try and fit the results by a model which takes into account both absorption and Fresnel near-field diffraction by the cavity. A possible setup is shown in Fig. 14.3.

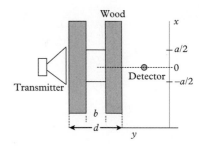

Figure 14.3 *Defect detection in wood.*

The Schottky diode detector is in this case stripped of its receiving horn, so that one can scan the intensity carefully along the x-axis on either side of the center of the hole in order to gain information about a, and along the y direction for b.

A possible model function for the interpretation of the scattered intensity would be a combination of exponential attenuation $\exp(-\alpha x)$ for absorption and of the Fresnel–Cornu integrals $I_{\text{F–C}}$ for diffraction by a hole of given dimension:

$$I(x,y) = I_{\text{F–C}}(x,y)e^{-\alpha(d-b)}, \quad |x| \leq a/2$$
$$I(x,y) = I_{\text{F–C}}(x,y)e^{-\alpha(d-b)} + I_0 e^{-\alpha d}, \quad |x| > a/2,$$

in which α is the coefficient of absorption and I_0 is the intensity without the wood. Measurements for different-sized holes, say for a in the range 2–5 cm and b in the range 1–6 cm, would validate and set limits for the model for the interpretation of the scanned intensity profiles.

REFERENCES

1. A. Lord, R.M. Koerner, J.E. Brugger, J.W. Beck, and D.C. Maricovitch, Moisture measurements in sand using a basic microwave setup, *Am. J. Phys.* **45**, 88 (1977).
2. M.T. Martin, E. Manrique, and A. Fernandez, An experiment with microwave radiation to determine absorption coefficients and defects in wood, *Phys. Edu.* **28**, 386 (1993).
3. N. Lindgren, PhD thesis, 2007, Microwave sensors for scanning of sawn timber, available online at <http://pure.ltu.se/portal/files/489731/LTU-DT-0709-SE.pdf>.
4. Johannes Welling, Moisture content in industrial wood processing, available online at <http://www.coste53.net/downloads/sopron/CostE53_Sopron-Welling.pdf>.

The Doppler effect

15

Introduction

The shift in frequency observed for all waves when source and detector are in relative motion has become a widespread tool in both research and engineering applications whenever detailed knowledge of motion is required. Employing both sound and electromagnetic waves, uses of the Doppler shift range from astronomy, meteorology, and medicine to police and military uses, as well as non-invasive vibration measurements in engineering and biological structures via laser Doppler interferometry [1]. The aim of this project is to become familiar with the Doppler shift through the use of microwaves and to apply it to simple velocity measurements.

Introduction	89
Theoretical ideas	89
Goals and possibilities	90
Experimental suggestions	90

Theoretical ideas

Sound waves

Let f be the frequency of an emitter in a frame at rest relative to it. Suppose both source and observer are moving through the transmitting medium, with velocities v_s and v_o, respectively, and let the speed of sound waves be v. Then the frequency f' detected by the observer is given by

$$f' = f\left(\frac{v \pm v_o}{v \mp v_s}\right), \tag{15.1}$$

where the upper signs are relevant when source and observer approach along the line joining the two, and the lower signs when they move apart.

If the source moves away from the observer and towards a reflector, or if a wave is reflected from a moving object, the superposition of the direct wave and the reflected wave results in beats whose frequency can be used to determine the speed of the moving source or object.

Electromagnetic waves

Here, (15.1) needs to be modified by relativity theory. We just quote the relevant formulas; their proofs may be found, for example, in Refs [2,3].

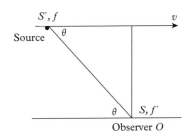

Figure 15.1 *Moving observer.*

Let a source in frame S emit waves of frequency f, and let an observer be in frame S' moving with velocity v relative to S. Then the frequency observed in S' is given by

$$f' = \left(\frac{1 \mp \beta}{1 \pm \beta}\right)^{1/2} f, \quad (15.2)$$

where $\beta = v/c$, and the upper and lower signs correspond to the cases of S' moving away from and toward S, respectively.

Suppose now that a source emits a signal of frequency f in its frame S', at an angle θ to its line of motion, moving with velocity v with respect to an observer O in S, as shown in Fig. 15.1. Then the frequency f' registered by a stationary observer at O will be given by

$$f' = \frac{(1-\beta^2)^{1/2}}{1 - \beta \cos\theta} f. \quad (15.3)$$

Both f' and the angle θ will be functions of time t. This formula has been used in tracking artificial satellites.

Goals and possibilities

- Use a standard demonstration microwave apparatus with a suitable experimental setup in order to test formulas (15.2) and (15.3).
- If an ultrasonic kit is available, use it to test (15.1) under various experimental circumstances.
- Use the above theory and suitably constructed apparatus to determine the velocity of various objects, such as the terminal speed of a spherical weight falling in oil or corn syrup (at least five diameters from the container walls), or a cart moving at constant speed. Compare the accuracy and precision of Doppler methods to alternative methods for finding these speeds.
- Determine the revolutions per minute (RPM) of a metal-blade rotating fan, or the frequency of a driven loudspeaker, by reflecting microwaves off them and using beat phenomena. Note that you can also measure RPM by analyzing the sound from a blade, or the sound reflected from the blade.
- Use interference of laser beams reflected from stationary and moving mirrors to generate beat patterns and from these determine the velocity of the mirror.

Experimental suggestions

At the basis of all the suggested experiments is the mixing of direct and reflected waves, displaying the resulting beat pattern on an oscilloscope, and relating the beat frequency to the velocity which is to be determined. Very useful texts on the Doppler effect and laser Doppler techniques are Refs [4,5].

Speed of rotation of fan

Use an adjustable auto-transformer to power a fan with metal blades (or plastic ones covered with aluminum foil) at different speeds. Measure the speed of rotation by using a stroboscope. Then direct a microwave beam through the gaps in the fan, perpendicular to the plane of rotation. Let a fan having n blades turn at a rate of N revolutions per second. Then the detector will register Doppler-shifted pulses with a repetition rate of nN.

Velocity measurement

Figure 15.2 shows a useful setup. A metal reflector is attached to a rider on an air track which can be moved at constant speed by a suitable motor, pulley, and weight combination. A stationary reflector is positioned near one end of the air track. Microwaves from an emitter are reflected from both reflectors, interfere with slightly different frequencies ($\beta \ll 1$), and produce beats. These are detected by the receiver and can be displayed on an oscilloscope. The beat frequency can be determined from the trace, and from it the velocity of the rider may be calculated. It may be compared with that measured by suitably positioned photogates.

An alternative is to replace the pulley–motor arrangement with a weight falling with constant terminal velocity in mineral oil attached to a pulley–cart combination.

Measuring loudspeaker vibration

Affix a sheet of aluminized mylar to the moving cone of a loudspeaker driven by a signal generator. Here we have an oscillating reflector, as well as a stationary reflector of microwaves from a source, as shown in Fig. 15.3. It is useful to restrict frequencies to the band 10–600 Hz. Here one has superposition of a wave from the stationary reflector and a Doppler-shifted wave from the mylar.

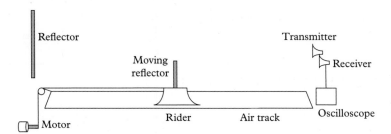

Figure 15.2 *Doppler apparatus.*

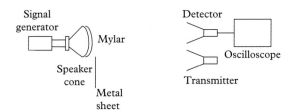

Figure 15.3 *Oscillating and stationary reflectors.*

The low-frequency envelope from the detector is displayed on the oscilloscope. One obtains a phase modulated beat frequency, that is, it varies sinusoidally, the periodicity of the beat frequency matching that of the source.

Angle dependence of Doppler shift

Rather than placing transmitter and detector at the end of the air track, as Fig. 15.2, place them at some distance at right angles to the track. Then, as the rider moves it will make varying angles with the source beam. The reflectors are now turned at right angles to their position in Fig. 15.2. One can then try to monitor the beat frequency as a function of the angle, which should agree with results from (15.3).

Velocity measurement with a laser

The air track setup may again be modified in order to see the beat phenomenon with a laser beam, rather than microwaves. Split a laser beam with a half-silvered mirror mounted at one end of the air track. One part is reflected from the mirror mounted on the moving rider. The other is reflected from a stationary mirror. The two beams are combined and focused by a lens along the track onto a photodiode whose output will display the beat frequency. The signal should be low-pass

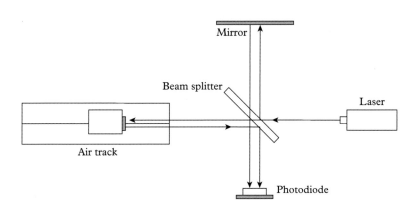

Figure 15.4 *Laser Doppler (Michelson) interferometer.*

filtered to eliminate frequencies higher than $f-f'$. Noise and vibration in the equipment may necessitate the use of a lock-in amplifier to pick out the signal, typically of the order of 1.5 MHz. The velocity of the moving cart on which the mirror is mounted can then be calculated from the relativistic Doppler shift formula and the result obtained checked against a photogate determination of the velocity. The setup is shown in Fig.15.4.

....

REFERENCES

1. R. Traynor, From the big bang to cleaner teeth: Doppler shift – a practical engineering tool, *J. Phys.: Conf Ser.* **105**, 012005 (2008).
2. A.P. French, *Special Relativity* (Nelson, London, 1968), pp. 137, 147.
3. E.F. Taylor and J.A. Wheeler, *Spacetime Physics* (W.H. Freeman, New York, 1992), p. 263.
4. L.E. Drain, *The Laser Doppler Technique* (Wiley, New York, 1980).
5. T.P. Gill, *The Doppler Effect* (Logos Press, London, 1965).

16 Noise

Introduction 94
Theoretical Ideas 94
Goals and possibilities 97
Experimental suggestions 97
Suggested investigations 99

Introduction

Many signals of scientific or technical interest are small and so can be significantly *altered* by either extrinsic noise ("pickup") or by intrinsic noise mechanisms present in the measurement chain itself. Because of this, we will explicitly represent our measured voltages as the sum of the desired signal and an unwelcome noise signal:

$$V(t) = V_0(t) + V_{\text{noise}}(t).$$

In this project, continuous random electrical noise is generated using reverse-biased Zener diodes, and analyzed by both analog and digital methods. Experiments with narrow- and broad-band noise, some purely electrical and some acoustic, with one and two noise sources, are described. Besides fast Fourier transform (FFT) computer programs for spectral analysis, various analog methods can be used. A considerable amount of simple electronic construction is involved. The experimental methods suggested aim to enhance understanding, rather than achieve state-of-the-art skills. Many other sources of noise, whether continuously generated or recorded, could be analyzed with the same techniques.

Theoretical Ideas

Time-varying electrical signals may either be described in the time domain, characterized by amplitude (in volts or watts) as a function of time, or in the frequency domain, by the spectrum, which is a plot with frequency along the horizontal axis and voltage or power on the vertical axis.

A pure, sinusoidal signal $V_0(t) = V_0 \sin(\omega t)$ has a root mean square (RMS) value of the voltage given by $V_{\text{0rms}} = \sqrt{(2)} V_0/2 = 0.707 V_0$, which is a useful measure of the spread of the signal values about the mean. The spectrum of a single-frequency sinusoidal signal of infinite duration is a narrow line at that frequency, with width determined by frequency stability. The height of the line is proportional to the amplitude.

On the other hand, any real signal is also characterized by noise in its amplitude distribution, $V_{\text{noise}}(t)$, characterized by its own RMS spread. The amplitude distribution of most random noise is Gaussian, meaning that any particular voltage V has a probability

$$p = [1/(V_{rms}\sqrt{(2\pi)}]\exp[-(V^2/2V_{rms}^2)]. \qquad (16.1)$$

Table 16.1 *Gaussian noise distribution.*

V	0	0.1	0.5	1	2	3	4	5
p	0.400	0.398	0.353	0.243	0.054	0.004	$1.3 \cdot 10^{-4}$	$1.49 \cdot 10^{-6}$

Figure 16.1 illustrates a Gaussian probability distribution for the noise component of a signal, using the probabilities given in Table 16.1 for various values of V (assuming $V_{rms} = 1$). Note the fast fall-off beyond $V = 1$.

Even after appropriate shielding has reduced extrinsic noise pickup to minimal levels, two types of noise commonly remaining are thermal ("Johnson") noise and $1/f$ noise. Johnson noise is electronic noise due to the agitation of electrons that is intrinsic to the measurement temperature(s). It is a form of "white" noise (i.e., having a flat spectrum: equal noise powers over different frequency intervals). Again, Johnson noise is due to random (thermal) fluctuations in the signal, causing the data to be scattered in a Gaussian distribution. For a small resistance the intrinsic fluctuations *are themselves small*, but nevertheless constitute an intrinsic noise floor that cannot be overcome.

On the other hand, there are myriad possible causes of $1/f$ noise (also called "pink" noise). One prominent explanation is that in the conducting material, the atoms are densely packed together but, due to disorder, there are still "openings" where the atoms can move. As resistance in a metal *arises* from disorder in the atomic lattice, changes in atomic positions will change the scattering of electrons traversing the sample. If two positions for an atom are equally favorable, then the atom can oscillate between these two positions, though the presence of neighboring atoms may create some barrier to this motion. Reasonable models of disordered systems (e.g., due to Bert Halperin and others) easily reproduce $1/f$ noise as arising from a simple distribution of barrier heights between atomic positions. At low signal frequencies (and sufficiently low signal voltage amplitudes), these oscillations can dominate the noise associated with a signal, and as the frequency decreases further, this effect becomes more and more pronounced.

Spectral analysis by analog methods can be done in several ways:

1. Narrow-band tunable filters can be used over a limited range.

2. *Add*, linearly, a sinusoidal signal (from an adjustable frequency signal generator) to the noise signal to be analyzed, so that beats form, which can be measured after passing through a low-pass filter. By measuring the amplitude of the beat signal as a function of the frequency of the oscillator one can perform a spectral analysis. (The noise must be bandwidth limited so as not to have components in the range of the low-pass filter.)

3. The signal to be analyzed is *multiplied* by a reference signal of adjustable frequency. From a trigonometric identity for the multiplication of

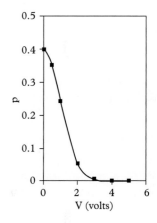

Figure 16.1 *Probability curve.*

two sinusoids, we expect to generate components at the sum and difference of the frequencies of the signal and the reference. The amplified result is then fed to a fixed narrow-band filter (sometimes a crystal) and measured. Again, by measuring the amplitude of the result as a function of the frequency of the reference signal one can generate a spectral analysis. In practice, this is done by modulation in a mixer, as in AM superheterodyne radio or a lock-in amplifier.

4. Spectral analysis by digital methods can be done by feeding the noise to a computer containing an appropriate FFT program (e.g. iSpectrum, Dog Park Software Ltd, <http://www.dogparksoftware.com>; or try <http://www.fftw.org/links.html>) yielding beautiful results. *Nota bene*: the user will not appreciate how results were obtained without personally writing an FFT program.

For a noise signal, the bandwidth Δf determines a coherence length, L_{coh}. The concept of a coherence length may be familiar from optics or acoustics, as the distance that separates detectors in a wave field so that their outputs no longer interfere, whether adding or canceling. $L_{coh} = c/\Delta f$, where c is the speed of the wave and Δf is the bandwidth. For a sinusoidal wave the coherence length is very large, but for a noise signal it is short enough to play a critical role. We expect this intuitively, since $1/\Delta f$ is a characteristic time for things to change and $c/\Delta f$ is the corresponding distance for loss of coherence.

A Zener diode conducts in the forward direction like any diode and only conducts or "breaks down" in the reverse direction when a particular voltage is reached, making it a useful voltage regulator. Electrons and holes in this "avalanche breakdown" generate a randomly fluctuating current and voltage with a spectrum that is flat from audio to microwaves, making it a convenient noise

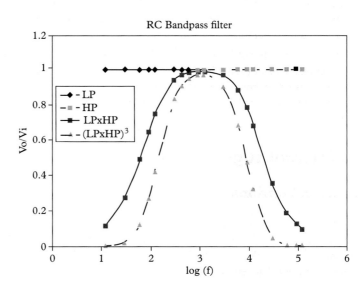

Figure 16.2 *Frequency response for various filters.*

source for many experiments. In this project filters are used to limit the bandwidth to audible frequencies between 100 and 10,000 Hz.

As an example, Fig. 16.2 shows the calculated response, namely voltage out, divided by voltage in, V_o/V_i, as a function of frequency f, for the bandwidth-limiting filters and bandpass filters of Fig. 16.4. Plotted are the responses of single low- and high-pass elements separately (LP V_o/V_i) and together (LP × HP), as well as that of three stages of each, effectively in series, (LP × HP)^3. Response curves for the filters shown in Fig. 16.6 can be calculated and in all cases they should be measured using an appropriate signal generator and voltmeter.

Before starting you may want to look at definitions of noise in a text [1], and online in Wikipedia.

Goals and possibilities

- Familiarize yourself with noise signals, what they look and sound like, through various bandwidths, from narrow to wide.
- Measure the RMS value of a given noise signal and its amplitude distribution and spectrum.
- Observe beats through a low-pass filter between a sinusoid from a signal generator and noise (this is a primitive form of spectral analysis).
- With amplified noise coming from one loudspeaker observe coherence length using two microphones, one stationary and the other movable, with amplified outputs displayed as X versus Y on the oscilloscope.
- With two speakers separated by some reasonable distance (0.3 to 1 m) and one movable microphone on a horizontal boom that can rotate about a vertical pivot halfway between the speakers, observe interference in the sound field with (a) a sound source of ∼1 kHz; (b) noise of various bandwidths from a single noise generator; (c) noise of various bandwidths from two separate noise generators.
- A very large number of further related projects can be considered containing combinations of topics, here indicated by just a few words: signals and noise (phase detection, auto- and cross-correlation), shot noise, Johnson noise, everyday noises like rain, hail, or sandstorm, signals from radioactive decay.

Experimental suggestions

Commercial options

- Audio signal generator capable of producing either sine or square waves (10 Hz to 100 kHz; inexpensive function generators perform badly at lower frequencies, so exploration of $1/f$ noise benefits from the availability of a higher-end signal generator).

- Amplifiers and speakers (add-on or low-cost stereo).
- Digital oscilloscope with memory/FFT.
- Use of a lock-in amplifier for this purpose is particularly well described by Libbrecht et al. [2].

Homemade options

Detailed explanations of the circuits shown below would take up an inordinate amount of space, but we refer the interested reader to Refs [1–5]. Note that capacitance values are in microfarads unless labeled otherwise.

1. Two noise generators, shown in Fig. 16.3.
2. A bandwidth-limiting RC filter (100 Hz to 10 kHz), as in Fig. 16.4.
3. A low-pass filter (0 to 10 Hz), shown in Fig. 16.5.
4. A tunable filter from an adjustable RC tee in the negative feedback loop of an op-amp, using the circuit in Fig. 16.4.
5. An adjustable parallel RLC filter driven by a transistor amplifier followed by an emitter follower, as shown in Fig. 16.6, where $f_{notch} = 1/2\pi C(3R_3R_5)^{1/2}$ for 25 to 100 Hz, $C = 0.1$, $R_1 = 1K$, $R_2 = 20K$, $R_3 = 60.4K$, $R_4 = 50K$, $R_5 = 464K$, $R_6 = 500K$.
6. Two microphone amplifiers capable of power output ~ 1 W. These are also used to drive miniature lamps for the RMS measurements. All these devices should be tested or calibrated to be sure that they conform to specifications.

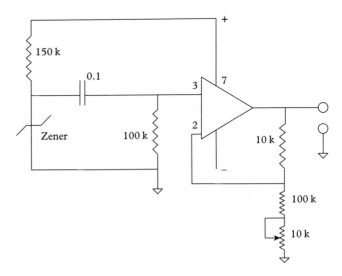

Figure 16.3 *Noise generator circuit.*

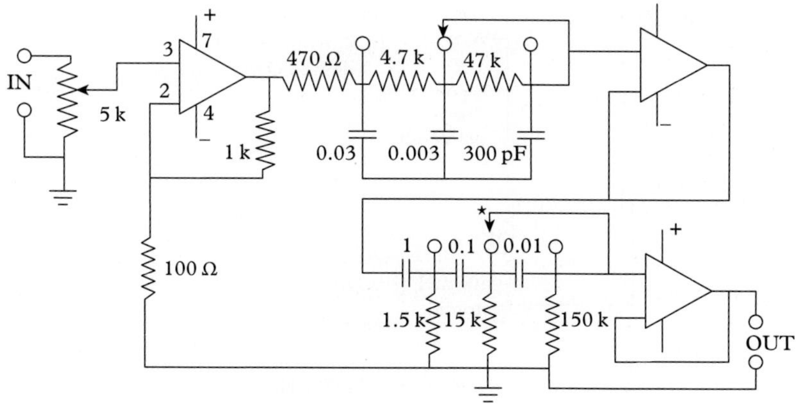

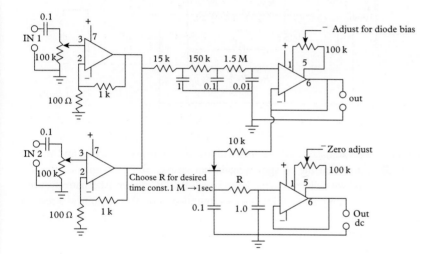

*2 optional switches (or clip leads) to choose fall-off

Figure 16.4 *Bandwidth limiting filter.*

Figure 16.5 *Low-pass filter and rectifier.*

Suggested investigations

Look at the noise signal from one of the generators directly (without filters) on the oscilloscope at various sweep speeds, slowest to fastest.

Connect the noise output of one of the microphone preamplifiers to a miniature flashlight lamp so as to light it as in normal operation. This lamp is mounted facing another about 0.1 m away that is supplied with rated DC voltage to light it. Place a piece of paper with a spot of thin oil half way between the lamps (this neat trick is called a Bunsen oil spot photometer): when the light outputs match, the oil spot, viewed in a darkened room, will not be visible. Now switch the lamp lit

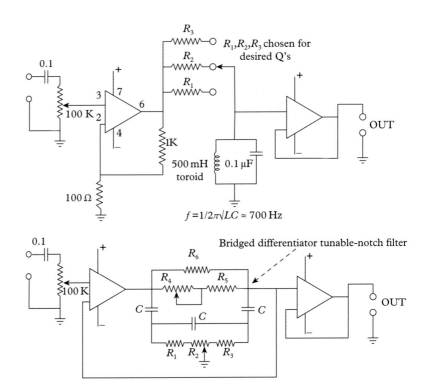

Figure 16.6 *Filter circuits.*

by noise to an adjustable DC supply with meters to measure current and voltage and adjust it so that the oil spot is again not visible. You can conclude that the RMS noise power delivered to the lamp equals the DC power. Alternatively, use a photodiode to compare light intensities (amplified noise also could heat a resistor in contact with a thermistor, calibrated with DC). By consulting the literature, compare this method with analog electronic and digital methods of determining RMS power.

The use of various filters is suggested in what follows. In each case after construction their frequency response should be measured using an appropriate signal generator and voltmeter. Also measure the dynamic range, noting the smallest sinusoidal signals that can be seen on an oscilloscope and the largest that show negligible distortion, typically a flattening of the peaks of the sine wave.

Look at the noise signal from one of the generators on the oscilloscope through the bandwidth-limiting filter at various sweep speeds. Find the moderately fast sweep time setting for which you can see about 30 plus and minus peaks on the screen, the largest extending to full screen, and either store (with a storage scope) or photograph so that you can count the number of peaks within, say, a 1 mm interval on the vertical scale. Repeat this measurement a few times and make an amplitude distribution plot.

With the oscilloscope in its X-Y mode, observe first one noise signal directly on X and Y, and then the signal from two independent noise generators, one on X and the other on Y.

Add two sinusoidal signals linearly (one from a signal generator, the other a low voltage at mains/line frequency) and observe the beats through the low-pass filter. Then add a sinusoidal signal (from a signal generator) to noise of various bandwidths, limited and narrow, and observe the beats as DC after rectification and filtering.

The coherence length is determined by measurements of the noise sound field from one speaker with two microphones initially side by side. Compare the two microphone outputs, one on the X and the other on the Y input. We expect a noisy straight line at 45°, and as one microphone is moved back away from the source and the coherence length is reached, the scope pattern should turn into a square full of noise as seen previously with two separate noise generators. Try this with different bandwidths. For the bandwidth-limiting filter, $\Delta f = 9900$ Hz and, with $c = 340$ m/s, $L_{coh} = 34$ mm (very small); with an RLC filter with $Q = 50$ at 1 kHz, $\Delta f = 20$ Hz and $L_{coh} = 17$ m (quite large).

....

REFERENCES

1. A. van der Ziel, *Noise in Measurements* (Wiley, New York, 1976).
2. K.G. Libbrecht, E.D. Black, and C.M. Hirata, A basic lock-in amplifier experiment for the undergraduate laboratory, *Am. J. Phys.* **71**, 1208 (2003).
3. P. Horowitz and W. Hill, *The Art of Electronics* (Cambridge University Press, Cambridge, 1990), and later editions.
4. R.F. Graf, *Encyclopedia of Electronic Circuits* (BPB Publishers, India, 2006).
5. E.B. Wilson, *Introduction to Scientific Research* (Dover, New York, 1990).

17 Johnson noise

Introduction

This project is a continuation of the previous one. Its importance lies in the fact that any estimate of the limiting signal-to-noise ratio of an experimental apparatus must take into account thermal (or Johnson) noise. The project focuses on measuring the voltage across a resistor caused by this noise and is an exercise in the choice of the right instruments and components to measure a very small effect.

Theoretical ideas

The fluctuating current due to the random thermal motion of electrons in a resistor gives rise to an effective fluctuating EMF in the resistor. The Nyquist theorem states that the mean square voltage across a resistor of resistance R in thermal equilibrium at temperature T is given by

$$\langle V^2 \rangle = 4R k_B T \Delta f, \quad (17.1)$$

where Δf is the frequency range within which the voltage fluctuations are measured, and k_B is Boltzmann's constant. For example, the root mean square voltage at room temperature for a $10\,k\Omega$ resistor and $10\,kHz$ bandwidth is $1.4 \cdot 10^{-6}$ volts.

This voltage does not depend on the details of the conductivity mechanism and so it may be derived by using either the Maxwellian or the Fermi–Dirac distribution of electron velocities. Equation (17.1) shows that noise may be reduced by lowering the temperature. Johnson showed experimentally the linear relation between $\langle V^2 \rangle$ and R. Thorough discussions, applications, and derivations of the theorem may be found in Refs [1,2].

Experimental suggestions

Low-noise resistors are a must. As we associate $1/f$ noise with disorder, this can be minimized by avoiding carbon composite/thick-film resistors and instead using metal thin-film resistors. They must be mounted in a screened box, to eliminate pickup. The apparatus needed includes a signal generator and attenuator, an amplifier with a gain of the order of $5 \cdot 10^3$, a frequency meter, oscilloscope, and a filter in the frequency band 5–20 kHz.

The required amplifier gain may be estimated as follows. Assume $\Delta f = 10\,\text{kHz}$, a 1 k$\Omega$ resistor and say 0.5 mV detection on an oscilloscope. By (17.1), this demands a gain of at least 1200. One needs a low noise and stable amplifier. If a digital oscilloscope is used to detect the signal, make sure that its timebase is such that its bandwidth is larger than that of the amplifier. This is achieved by putting a bandwidth filter of 5–20 kHz on the amplifier.

The circuit diagram in Fig. 17.1 shows a low-noise two-stage amplifier [3]. The first stage uses a low-noise HA5170 op-amp to provide appropriate input characteristics in a low-gain amplifier, while the second stage uses an HA5147 and provides much higher gain. Typical values for the resistors are $R_1 = 10\,\Omega$, $R_2 = 100\,\Omega$, $R_3 = 2.2\,\text{k}\Omega$. The gain of the first stage is 11, and the second stage 221. The op-amps are powered by 12 V levels applied in parallel with 0.1 μF capacitors to ground, to filter off noise in the power supply. As shown in the block diagram of Fig. 17.2, a voltage divider can be used as an attenuator on the output of the signal generator, in order to avoid saturating the amplifier.

If a resistor R is connected across the input terminals of an amplifier whose frequency-dependent voltage gain is $G(f)$, then the noise voltage appearing at the output of the amplifier, $\langle V_o^2 \rangle$, is given by [4]:

$$\langle V_o^2 \rangle = \langle V_a^2 \rangle + 4kTR \int_0^\infty [G(f)]^2 df, \qquad (17.2)$$

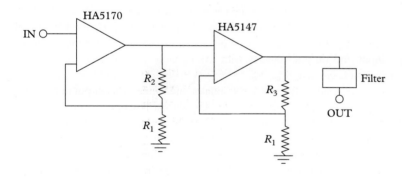

Figure 17.1 *Circuit for the two-stage amplifier.*

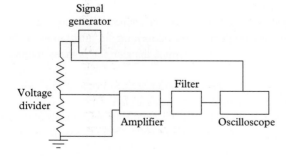

Figure 17.2 *Setup for measuring the gain $G(f)$.*

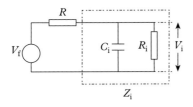

Figure 17.3 *Equivalent circuit.*

where $\langle V_a^2 \rangle$ represents the mean square noise voltage across the output terminals generated by the amplifier, and the second term is the noise voltage from the resistor.

Equation (17.1) describes the thermal noise voltage V_f associated with a resistance R, which Fig. 17.3 schematically represents as an alternating voltage generator V_f driving a current I through the resistance R, which in turn is connected to the *input impedance* of the amplifier. In general, we must consider how each subsequent stage in a measurement "chain" will affect (or "load") the input from the previous stage. Here, to assess how the input circuit of the amplifier modifies the signal, we must assess the input impedance of that stage. We do this by modeling that input circuit as a simpler, equivalent circuit of impedance Z_i produced by the parallel combination of the effective input capacitance C_i and the effective input resistance R_i. Note that here, instead of the voltage V_f, it is the voltage V_i that appears at the input to the amplifier.

Straightforward analysis of the schematic in Fig. 17.3 (which acts as a frequency-dependent voltage divider) shows that the second term on the right-hand side of (17.2) must be modified by dividing it by a factor A, where

$$A = (1 + R/R_i)^2 + (\omega C_i R)^2. \tag{17.3}$$

The values of R_i, ω, and C_i are characteristics of the op-amps (and C_i must also include the capacitance of the coaxial cables used). The correction A is less than 1% for $R < 30\,\text{k}\Omega$.

Before measuring noise voltages it is first necessary to calibrate the amplifier and determine $G(f)$. First calibrate the attenuator. Disconnect it from the amplifier and, using a sine wave, apply a signal to the attenuator input. Measure the input and output voltages with the oscilloscope and hence calculate the attenuation coefficient, $V_{\text{out}}/V_{\text{in}}$.

Now feed a sine wave into the attenuator and connect the output of the attenuator to the input of the tuned amplifier. Use the oscilloscope to monitor the input signal to the attenuator and the output from the amplifier. Tune the signal generator to the center frequency of the system, and measure the gain. Do not forget that the attenuator is in the circuit and that the signal voltage at the input of the amplifier is the signal from the oscillator reduced by the attenuation coefficient. By altering the frequency of the signal generator, determine the system gain over its useful frequency range.

The integral in (17.3) can now be determined by plotting a graph of the square of the gain against frequency and performing an integration of the area under the curve. To reduce errors, determine the gain for as many frequency values as possible.

The quantity $\langle V_a^2 \rangle^{1/2}$ may be found by measuring $\langle V_o^2 \rangle^{1/2}$ when the input is shorted with a copper wire. Then using a short length of coaxial cable one can connect different resistors, in turn, across the input terminals of the amplifier and measure the output voltage. A plot of $\langle V_o^2 \rangle - \langle V_a^2 \rangle$ versus R should be a straight line whose slope will give k_B.

Noise voltage as a function of temperature may be investigated by suitably immersing the resistors in liquid nitrogen, dry ice, or in metal tubes heated to different temperatures.

REFERENCES

1. C. Kittel, *Elementary Statistical Physics* (Wiley, New York, 1958).
2. F. Reif, *Fundamentals of Statistical and Thermal Physics* (McGraw-Hill, New York, 1965).
3. J. Napolitano, *Laboratory Notes for Physics 2350*, Rensselaer Polytechnic Institute, 1999.
4. The University of Sheffield, Department of Physics, *Johnson Noise E17*, 2006; see also their website, <https://www.sheffield.ac.uk/physics>.

18 Network analogue for lattice dynamics

Introduction 106
Theoretical ideas 106
Goals and possibilities 109
Experimental suggestions 109

Introduction

By making analogies to electrical networks, problems in many fields have been simulated and solved, including complex phenomena associated with fluid dynamics, diffusion, acoustic systems, and wave and transport phenomena. A key point is that electrical networks (comprising resistances, inductances, and/or capacitors) are relatively easy to construct, adjust, and analyze – all of which allows for a wide range of systematic studies. As an introduction to these methods this project makes use of an electrical network to simulate the thermal properties of a solid material. In particular, we wish to explore the vibration spectrum of a linear monatomic and diatomic lattice, and of a lattice with an impurity. For these topics in lattice dynamics no direct work at the atomic level is available outside of the most advanced laboratory settings. Nevertheless, the analogue system described here (and based entirely on [1]) should be useful to a student when studying the spectrum of lattice vibrations ("phonons") in either a statistical physics course or a solid state physics course.

Theoretical ideas

In thinking about waves in general, a great deal of attention is given to the relationship between energy and momentum, which is often expressed in terms of the relationship between frequency and wavenumber, $\omega(k)$. These relationships determine a great deal about wave behavior including, for example, how a traveling pulse will spread out as it propagates. Commonly referred to as the "dispersion relationship," $\omega(k)$ plays a central role in any solid state physics course, where it is common to develop, as specific examples, the dispersion relations for linear chains of monatomic and diatomic lattices, and to consider the localized modes that are induced by the addition of impurity atoms [2].

The basis for the simulation is that the mathematics turn out to be equivalent to the theory of the electrical and mechanical units shown in Fig. 18.1. The dynamic equations for I and x are

$$M\ddot{x} + 4fx = 0; \quad L\ddot{I} + \frac{4}{C}I = 0. \tag{18.1}$$

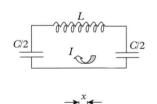

Figure 18.1 *Electrical and mechanical equivalents.*

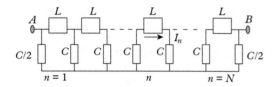

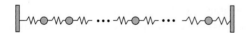

Figure 18.2 Linear electrical "lattice array".

The pairs (I, x), (L, M), and $(1/C, f)$ are corresponding quantities in the analogy, where, in this notation, $2f$ is the spring constant.

A linear electrical array of such units will correspond to a monatomic lattice of finite extent, as shown in Fig. 18.2. A neat result is that for N coupled oscillators there are N allowed normal modes of oscillation for the system. For the electrical chain these modes are characterized by the currents:

$$I_n = A \sin \omega_r t \sin\left[\left(n - \frac{1}{2}\right) k_r\right], \qquad (18.2)$$

where $\omega_r = \omega_c \sin \frac{1}{2} k_r$ and $r = 1, 2, \ldots, N$.

The connection between the electrical and mechanical chains is established by requiring that $k_r = r\pi/N$ or, when expressed in terms of wavelengths, $\lambda_r = 2\pi/k_r$, that $r\lambda_r/2 = N$. Note that there is a minimum wavelength, which in the case of a lattice of atoms is determined by the lattice spacing. The associated maximum ("cut-off") frequency is given by $\omega_c = 2/(LC)^{1/2}$, and occurs when $k_r = k_N = \pi$.

If the terminals A and B are electrically connected, converting the linear array into a (topological) ring, this removes any endpoints from the simulation, making it more appropriate for simulating very large samples, where the peripheries are expected to have negligible effect upon bulk properties such as the spectrum of lattice vibrations. In this case, the current must satisfy the periodic boundary condition $I_{n+N} = I_n$. If N is even, there will then be $N/2$ distinct normal modes, satisfying $k_r = 2r\pi/N$ or $r\lambda_r = N$, with $r = 0, 1, 2, \ldots, N/2$. The currents will then be given by

$$I_n = A \sin \omega_r t \sin[nk_r + \varphi]. \qquad (18.3)$$

The phase angle φ will be determined by the particulars of the coupling to the signal generator and by choosing for a particular unit $n = 1$.

Arrays may also be constructed to represent a diatomic lattice or impurities. For the diatomic lattice one has alternate units in the chain having inductances L_1, L_2. The unit cell will contain a neighboring pair of these oscillators. For an even number N of such oscillators, closed in a ring, the normal mode currents will be given by

$$I_{2n} = A \sin \omega_r t \sin(2nk_r + \varphi), \quad n = 1, 2, \ldots, N,$$
$$I_{2n-1} = B \sin \omega_r t \sin[(2n-1)k_r + \varphi], \quad (18.4)$$

where $k_r = r\pi/N$, $r = 1, 2, \ldots, N/2$. L_1 and L_2 are associated with cell numbers $2n$, $2n-1$, and amplitudes A, B, respectively.

Each value of k_r has two modes associated with it, with frequencies given by

$$\omega_r^2 = (\omega_1^2 + \omega_2^2)\left[1 \pm \left(1 - 4\frac{\omega_1^2 \omega_2^2}{\omega_1^2 + \omega_2^2}\sin^2 k_r\right)^{1/2}\right], \quad (18.5)$$

where $\omega_1^2 = 1/L_1 C$, $\omega_2^2 = 1/L_2 C$, and

$$\frac{A}{B} = \frac{2\cos k_r}{\omega_r^2/\omega_1^2 - 2} = -\frac{\omega_r^2/\omega_2^2 - 2}{2\cos k_r}.$$

The presence of a light impurity in the pure lattice is known to give rise to a localized mode having a frequency which is above the cut-off value ω_c of the regular array. In order to simulate a lattice impurity atom, insert a unit comprising an inductance L' and a capacitance $C'/2$ in the above electrical network. In order to calculate the frequency of the localized mode, use is made of the concept of characteristic impedance Z_0. This is defined for a single unit in Fig. 18.3: When Z_0 is connected across the output of a unit, the input impedance also equals Z_0. The input impedance of a linear array terminated by Z_0 is also Z_0. One finds from Fig. 18.3 that

$$Z_0^2 = \frac{L/C}{1 - \omega^2/\omega_c^2}. \quad (18.6)$$

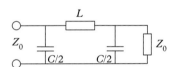

Figure 18.3 *Unit for localized mode.*

For frequencies $\omega > \omega_c$, the insertion into an infinite linear array of the "impurity unit" shown in Fig. 18.4, having its own resonance frequency $\omega_0 = 2/(L'C')^{1/2}$, is equivalent to the circuit shown in Fig. 18.5, where now $C_0 = \frac{1}{2}C(1 - \omega_c^2/\omega^2)^{1/2}$. The local mode frequency is given by

$$\omega_L^2 = 2 \bigg/ L'\left(C_0 + \frac{1}{2}C'\right), \quad (18.7)$$

where C_0 is evaluated at $\omega_0 = \omega L$. For the case that $C' = C$ (corresponding to the same force constant for the host and impurity atomic bonding) this becomes

$$\omega_L^2 = \omega_0^4/(2\omega_0^2 - \omega_c^2). \quad (18.8)$$

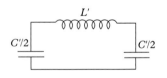

Figure 18.4 *Impurity.*

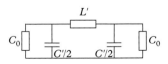

Figure 18.5 *Equivalent circuit.*

Since the disturbance caused by oscillation of an impurity unit above the cut-off frequency is sharply attenuated, in practice one needs to connect on either side of it only a few pure units.

Goals and possibilities

- Measure the dispersion relation $\omega(k)$ for a monatomic and diatomic lattice.
- For a diatomic array monitor the phases and amplitudes for "acoustic" and "optical" modes.
- Substitute several different unit oscillators with resonance frequencies above the cut-off into the array to simulate local modes, plot ω_L/ω_0 as function of ω_0/ω_c, and compare with (18.8).

Experimental suggestions

See Ref. [1]. Figure 18.6 shows the construction of the array. Prepare coil inductors from some 50 turns of 26 SWG (0.4 mm-diameter) enameled wire, tightly wound in a single layer on a 12 mm-diameter insulating rod. Its inductance is ~0.08 mH. Each coil has its own pair of capacitors $\frac{1}{2}C \approx 100\,\text{pF}$. The resonance frequency of a single isolated unit is ~5 MHz. An array of 12 such units is suitable for the simulation.

To simulate a diatomic system build six unit cells each having a pair of coils, with a 2:1 ratio of close-wound turns; each coil has a pair of capacitors $\frac{1}{2}C = 82\,\text{pF}$. Determine ω_1 and ω_2 in (18.5) by connecting to each type of inductor L_2 and L_1 a capacitor C of 164 pF and finding the resonance frequencies of the circuit. With $L_2/L_1 = 2.8$ these frequencies are 2.1 and 1.26 MHz, respectively.

Determining the relative sensitivities of the pick-up loops on the two types of coil is straightforward: just connect a separate specimen of each type of coil through a capacitor $C/2$ at each end to ground, drive this unit to resonance, and measure the induced voltages for each to give the required ratio.

Each such coil has a turn of wire wound on it as a pickup A, for measuring the current I_n. In addition, one coil in the array (in a ring configuration it does not matter which one) has a few turns on top of it for the input B, namely, coupling the array to a function generator.

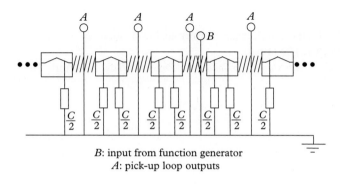

B: input from function generator
A: pick-up loop outputs

Figure 18.6 *Experimental array.*

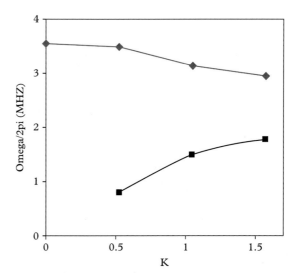

Figure 18.7 *Experimental dispersion relation for diatomic lattice (points) and theory (lines).*

Display the input as well as the voltages induced in the pickup loops on an oscilloscope with memory. Comparison with the input waveform enables one to determine the phases of the currents. The normal mode frequencies are located by watching for sudden increases of the oscilloscope wave amplitudes.

For studying local modes, prepare several units as in Fig. 18.4 with different resonance frequencies, all above the cut-off value, and insert each in turn into the center of a monatomic array of 12 units. This time drive the impurity unit from the function generator and couple it also to the oscilloscope in order to observe the resonance frequencies. Figure 18.7 shows the dispersion relation for a diatomic lattice [1], the points representing measurements, the curve calculated from (18.5).

....

REFERENCES

1. D.H. Tomlin, An electrical network analog of vibrations of a linear lattice, *Eur. J. Phys.* **1**, 129 (1980).
2. C. Kittel, *Introduction to Solid State Physics* (Wiley, New York, 2005).

Resistance networks

19

Introduction

Introduction	111
Theoretical ideas	111
Goals and possibilities	113
Experimental suggestions	113

Interconnected resistances arranged in ladder, square lattice, or toroidal network geometries represent interesting examples for researchers studying circuit theory, whereby Kirchoff rules and ideas of symmetry may be employed to carry out and simplify the analysis for such circuits. In the project researchers build finite networks, measure the values of effective resistances and voltages between any two nodal points, and carry out numerical solutions of the equations governing these quantities in order to check how these values approach asymptotically those obtained for infinite networks.

Theoretical ideas

Ladder network

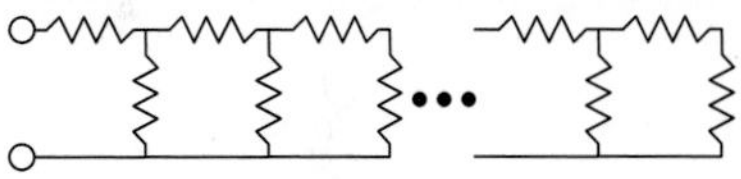

See Refs [1–4]. Suppose we have a ladder of n equal resistances R, as shown in Fig. 19.1(a).

Figure 19.1(a) *Resistance ladder.*

The equivalent resistance R_{n+1} across the terminals of the ladder with $n + 1$ loops can be found from the equivalent resistance R_n of n loops by referring to Fig. 19.1(b). The result is

$$R_{n+1} = \frac{R_0 + 2R_n}{R + R_n} R, n = 1, 2, 3 \ldots; R_1 = 2R. \tag{19.1}$$

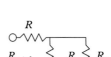

Figure 19.1(b) *Equivalent resistance.*

For an infinite network we have, setting $R_{n+1} = R_n = R_\infty$ and solving for R_∞,

$$R_\infty = \frac{1 + \sqrt{5}}{2} R. \tag{19.2}$$

It is of interest to investigate the rate at which R_n approaches R_∞. We may put $R_n = R_\infty + \delta_n$ in (19.1) and assume for n large enough that $\delta_n \ll R_\infty$. The resulting recursion relation, together with (19.2), leads to

$$\delta_{n+1} = 1/c^2, \; c = (3 + \sqrt{5})/2. \tag{19.3}$$

The decaying solution obeys $\delta_n \propto 1/c^{2n}$, so that R_n approaches R_∞ exponentially. Numerical calculation shows that already after two loops, $n = 2$, the relative deviation $(R_n - R_\infty)/R$ is of the order of 0.05.

If we apply a potential difference to the terminals of the ladder in Fig. 19.1(a), the currents in successive loops should also decrease exponentially. This may be shown by using Kirchoff loop currents i_k for the kth loop. For the infinite ladder ($n = \infty$) we find that $i_k \propto 1/c^k$ as $k \to \infty$. For finite ladders the most general solution that satisfies the recursion relation for the kth loop, namely, $3i_k = i_{k+1} + i_{k-1}$ with $k = 2, 3, \ldots, n-1$, is $i_k = Ac^{-k} + Bc^k$, where Kirchoff's rule for the final loop fixes $B/A = -c^{-2n}(3-c)/(3-1/c)$.

Two-dimensional grid

Consider the two-dimensional grid of resistances shown in Fig. 19.2. For details, see Refs [2–4].

Adjacent nodes

By symmetry arguments it follows that the resistance between adjacent nodes is $R/2$.

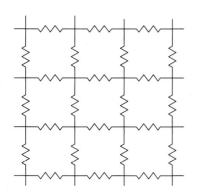

Figure 19.2 *Two-dimensional grid.*

Any two nodes

The resistance between any two points P and Q can be found from $2V/I$, where V is the voltage drop from P to Q when a current I enters at P and leaves at infinity.

Choose an arbitrary node as $(0,0)$ and denote the node in the mth row and nth column relative to it by (m,n), where n, m vary from $-\infty$ to $+\infty$. Then Kirchoff's rule for currents at (m,n) gives

$$I_{m,n} = (V_{m,n} - V_{m,n+1})/R + (V_{m,n} - V_{m,n-1})/R + (V_{m,n} - V_{m+1,n})/R + (V_{m,n} - V_{m-1,n})/R. \tag{19.4}$$

At a node where $I_{m,n}$ vanishes, (19.4) becomes

$$4V_{m,n} = V_{m,n+1} + V_{m,n-1} + V_{m+1,n} + V_{m-1,n}. \tag{19.5}$$

We assume that a current I enters at $(0,0)$ and leaves at infinity. Then

$$I_{m,n} = 0 \text{ and } I_{0,0} = 0. \tag{19.6}$$

The difference equations (19.4) and (19.5) may be solved by the method of separation of variables. We shall only quote the final results for $R_{m,n}$ and $V_{m,n}$, and refer the reader to Ref. [1] for details:

$$\begin{aligned} R_{m,n} &= (R/\pi) \int_0^\pi d\beta (1 - e^{-|m|\alpha} \cos n\beta)/\sinh \alpha \\ &= (R/\pi) \int_0^\pi d\beta (1 - e^{-|n|\alpha} \cos m\beta)/\sinh \alpha, \end{aligned} \tag{19.7}$$

$$V_{m,n} = (IR/\pi) \int_0^\pi d\beta (e^{-|m|\alpha} \cos n\beta + e^{-|n|\alpha} \cos m\beta)/\sinh \alpha, \tag{19.8}$$

where

$$\cosh\alpha + \cos\beta = 2. \tag{19.9}$$

The values for $R_{m,n}$ and $V_{m,n}$ may be found by numerical evaluation of these integrals. The values of $R_{m,n}$ for large m (or n) may be found from asymptotic expansions of the integrals in (19.7). We find [1,2] that

$$R_{m,n}/R = (1/2\pi)[\ln(n^2 + m^2 + 0.51469)]. \tag{19.10}$$

Actually, this expansion already gives good results for $R_{m,n}/R$ even for quite small m, n. For example, the values given for $R_{1,0}$, $R_{1,1}$, $R_{3,4}$, $R_{5,0}$, and $R_{10,10}$, are (0.5, 0.636, 1.028, 1.026, 1.358) and (0.515, 0.625, 1.027, 1.027, 1.358) by direct calculation from (19.7) and from (19.10), respectively.

It is remarkable [3,4] that for an infinite square grid the equivalent resistances across a side of the square, a diagonal, double side, knight's move, and double diagonal are given, respectively, by 0.5, $2/\pi$, $2 - 4/\pi - 0.5$, $8/3\pi$.

Goals and possibilities

- Build a ladder of resistances, successively adding loop after loop, so as to check the rate of convergence of the experimental data to R_∞.

- Investigate whether the end at which loops are added to the ladder influences the agreement between measurements of the effective resistance and the theoretical prediction. Loops may be added at the end where the multimeter is connected, or at the opposite end.

- Construct a 12×12 square grid of resistors and carry out measurements of effective resistances between nodes along a side, a diagonal, a double side, and double diagonal of the basic square unit, as well as nodes connected by a knight's move.

- Compare the measurements with the predictions of theory for an infinite grid.

Experimental suggestions

An accuracy of 1% in the resistors is desirable. Measure the resistances of all resistors used and using a histogram figure out their mean and standard deviation, so as to be able to assess the accuracy with which the measured and theoretical effective resistances agree with each other. For the ladder experiment the resistors may be mounted on an electronic breadboard, obviating the necessity for soldering.

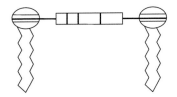

Figure 19.3 *Mounting method.*

Equivalent resistances should be measured by a multimeter of six-digit accuracy. For the square grid one may use nichrome wire looped around appropriately separated posts on a frame, silver soldering the overlapping points. An alternative, more accurate, method would be to mount a square grid of flat-top brass screws on a board, and solder (with a high-temperature iron) the resistor wires into the channels across the heads, as illustrated in Fig. 19.3.

REFERENCES

1. B. Denardo, J. Earwood, and V. Sazonova, Experiments with electrical resistive networks, *Am. J. Phys.* **67**, 981 (1999).
2. G. Venezian, On the resistance between two points on a grid, *Am. J. Phys.* **62**, 1000 (1994).
3. D. Atkinson and F.J. van Steenwijk, Infinite resistive lattices, *Am. J. Phys.* **67**, 486 (1999).
4. B. van der Pohl and H. Bremmer, *Operational Calculus*, 2nd ed. (Cambridge University Press, London, 1955), p. 371.

Part 3

Acoustics

20	Vibrating wires and strings	117
21	Physics with loudspeakers	124
22	Physics of the tuning fork	129
23	Acoustic resonance in pipes	134
24	Acoustic cavity resonators and filters	138
25	Room acoustics	141
26	Musical instruments: the violin	146
27	Musical instruments: the guitar	151
28	Brass musical instruments	155

Vibrating wires and strings

20

Introduction

The experiment on transverse vibrations in a string or wire is nowadays a standard laboratory exercise. We include it in this book because the variants we suggest lead to avenues not normally explored. Among these are resonance curves for particular modes, the dependence of the ability to excite various modes on the position of the driver and the detector, the change in mode structure when the string is compounded of two parts differing in density, the transition from the modes of a lumped to a continuous system, the overtone structure when plucking the string at various places, coupled piano strings, and the polarization of vibrations. There is enough exploratory material here for an extended project in depth, or for several shorter ones.

Introduction	117
Theoretical ideas	117
Goals and possibilities	119
Experimental suggestions	120

Theoretical ideas

We refer the reader to the book on waves in the Berkeley series by Crawford [1]. It is admirable in its treatment of all the topics that are relevant for this project, namely, resonance, standing waves, and beaded strings, as well as Fourier analysis of pulses. Much useful material may also be found in books by Morse and French [2–4]. References [5–9] treat the problems of interest in this project; we only summarize relevant results and refer the reader to the original articles for details.

First, we note that for a lightly damped sonometer string vibrating at small amplitudes A_n, the resonance frequencies of the nth mode f_n are multiples of the fundamental f_0, and because of energy conservation we expect that nA_n is a constant. Resonance curves have approximately the Lorentzian shape [1].

Secondly, we quote here the equations for a beaded string. Let beads of mass M be mounted at intervals d on a light string in equilibrium tension T_0. In a first approximation one neglects the mass of the string. Then the difference equation of motion for the transverse small displacement Ψ_n of the nth bead is [1]:

$$M\ddot{\Psi}_n = \frac{T_0}{d}\left[(\Psi_{n+1} - \Psi_n) - (\Psi_n - \Psi_{n-1})\right]. \tag{20.1}$$

Solution of these equations enables one to derive the dispersion relation connecting the angular frequency ω and the wavenumber k:

$$\omega = 2(T_0/\mu)^{1/2} \sin(k/2)/d, \tag{20.2}$$

where $\mu = M/d$ is the linear mass density of the beaded string. For a string of length $(N+1)d$ with N beads, the values of k are given by $k = \pi j/(N+1)d$, with $j = 1, 2, \ldots, N$.

We may take into account the finite mass of the string connecting the beads by assuming that for each section the mass of the uniform string is concentrated midway between the beads at $d/2$. We then have, effectively, a "diatomic" lattice with two masses M and $m = \mu d$, separated by a distance $d/2$. The dispersion relation is then modified to [5]:

$$\omega^2 = \frac{2T_0}{mMd}\left[(m+M) - \{(m+M)^2 - 4mM\sin^2(kd/2)\}^{1/2}\right]. \tag{20.3}$$

Thirdly, plucked or struck strings do not behave as the usual simple linear treatments would have us believe. The tension itself must vary periodically during displacement of the string, transverse vibrations are possible in two perpendicular directions (polarizations), and the vibrating string may be shown to be equivalent to two non-linear oscillators parametrically coupled. The result is the often-observed elliptically polarized motion precessing about the axis of symmetry. At large amplitudes the result may be a distorted non-symmetric resonance curve [6,7].

Plucked strings of a guitar or struck strings of a piano will, when excited in two perpendicular directions, produce sounds with different radiative properties, decay rates, and efficiencies, since the two modes couple differently to the soundboard.

Fourthly, consider the normal mode frequencies of a wire-wound guitar or piano string, some of whose length has been stripped of its windings. Let the two parts have lengths L_1, L_2, linear mass densities μ_1, μ_2, and wave velocities c_1, c_2. The velocities are calculated from $c = (T/\mu)^{1/2}$, where T is the tension. The normal mode frequencies f_n are given by solutions of the equation [8]

$$\cot(2\pi f L_1/c_1) = -(\mu_2/\mu_1)^{1/2}\cot(2\pi f L_2/c_2). \tag{20.4}$$

Fifth, stiff piano strings do not obey the usual rule that the partials are multiples of the fundamental [10]. This is because when the string is displaced a distance y at the position x, the restoring force is composed of two parts: the one usually treated, due to the tension T, which gives rise to a force equal to $T(d^2y/dx^2)$, and, in addition, an effect due to elastic stiffness, which yields a further restoring force of $ESK^2(d^4y/dx^4)$, where E is Young's modulus, S the cross-sectional area of the wire, and K its radius of gyration ($d/4$ for a round wire of diameter d). The allowed frequencies f_n depend somewhat on whether one assumes clamped or pinned boundary conditions for the piano strings, but real piano strings seem to be somewhere in between. For clamped strings of length L the frequencies are given by [8]

$$f_n = nf_0\{1 + (4/\pi)B^{1/2} + [(12/\pi^2) + n^2]B\}^{1/2}, \qquad (20.5)$$

where $f_0 = (1/2L)(T/\mu)^{1/2}$, and $B = (\pi^2 ESK^2/TL^2)$.

Sixth, the theory of coupled piano strings has been treated both theoretically and experimentally by Weinreich [9], and has also been discussed by Benade [11]. Several results are of interest. Many notes on the keyboard are generated by striking a triplet of wires. When a single wire of a piano triplet is sounded (the other two being clamped), and the time decay of the sound is then analyzed, it is found that the decay curve consists of two distinct sections, corresponding to different decay rates of the vertical and horizontal polarizations of the string vibration. These polarizations have different radiation efficiencies. Releasing another member of the triplet affects the decay curve profoundly.

Goals and possibilities

- Verify the tension dependence for the frequency of a particular mode for a horizontal stretched wire of given length fixed at both ends.
- Excite the wire electromagnetically (see experimental section) and measure the variation of the amplitude A_n as a function of mode number n for a given power input to the vibrating wire. Check the relationship $A_n n = $ constant.
- For some modes establish the resonance curve of power versus frequency and interpret the result.
- Study the effect of the position of the exciting magnet on the modes that can be excited and observed.
- Measure the normal mode frequency by optical means: stroboscopically, or by using a photointerruptor.
- Investigate the overtones generated when the string is plucked at various positions, and compare with theory. Plucking should be done in a controlled manner, for example by using a pendulum to collide with the wire.
- Compare the normal mode frequencies of a guitar string wire-wound along its entire length, with those when the wire winding is removed from part of the string.
- Waves excited on a stretched string are usually elliptically polarized. Also, the tension is not a constant due to the finite amplitude of the displacement. This can lead to non-linear effects and parametric excitations. Explore these effects.
- For a very light wire, string some beads at regularly spaced intervals and find experimentally the normal modes and the ω versus k dispersion relation; compare with theory and with the continuous string result. Vary the bead size and spacing.

- Analyze the normal mode frequencies of a stiff piano string in situ and compare with theory.
- Produce the decay curve of a piano triplet with two or one of the triplet clamped. Probe the polarization dependence of the decay curves and relate these to theoretical treatments.

Experimental suggestions

The equipment required consists of a signal generator able to vary the frequency in steps of 0.1 Hz, a data acquisition system, amplifier, thin light copper wire, beads of different sizes, wound guitar string, photointerruptor with suitable circuit and 5 V power supply, magnet (electromagnet or strong permanent one), small lapel microphone 1 cm in diameter, camera for recording amplitudes, and oscilloscope.

Amplitude versus mode number $A(n)$, and resonance curve $A^2(\omega)$

In order to investigate these topics the apparatus could be any variant of the usual sonometer setup, in which one end of the wire is clamped, the other passes over a pulley and weights may be hung on the wire end. An AC current with finely variable frequency (use an audio amplifier if necessary) is passed through the wire and the modes are excited by a magnet (permanent or electromagnet) placed across the wire and at various positions along it. The amplitudes at the antinodes may be recorded by a vertical scale placed nearby, or by photographing the vibrating wire with a fine grid behind it, and analyzing a video recording of the motion by a frame-grabber program.

Beaded strings

See Ref. [5]. A suitable setup for studying the beaded string is shown in Fig. 20.1. The knife edges are movable. To excite the modes proceed as in the previous experiment, by feeding a sinusoidal current from a finely tunable signal generator through the wire and placing a magnet near the end. Frequencies may be

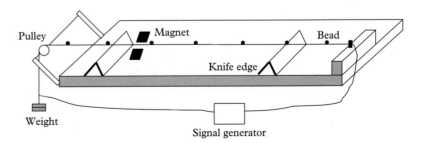

Figure 20.1 *Arrangement for beaded string experiment.*

measured with a frequency meter or by optical means: A commercially available photointerruptor (consisting of an infrared-emitting LED and a phototransistor, mounted in a moulded plastic frame with a 1 cm gap) is placed over the vibrating wire near one end. The device is connected to an oscilloscope which displays the waveform and measures the frequency. One edge of the wire is in the light beam and the other is outside. Very small string displacements may be measured this way. Alternatively, use a V-shaped opening which allows light from the illuminated string to reach a photodetector, the amount depending on the position of the string as it moves in the plane of the V.

SWG 34 copper wire may be used (400 mg per meter), strung with regularly spaced beads. One may prepare several differently beaded wires: smaller lighter beads more closely spaced, or heavier ones more sparsely distributed, keeping the average linear mass density the same, at say 10 g/m. This enables one to study the effect of inter-bead distance on the dispersion relations and on the cutoff frequencies.

Figure 20.2, reproduced from [5], presents dispersion relations for different beaded strings with the same mass density and tension, but different inter-bead separations a: curves 1, 2, 3, 4 correspond to $a = 25, 14, 10, 7$ cm, respectively. For comparison, the dotted line 5 shows $\omega(k)$ for a continuous string.

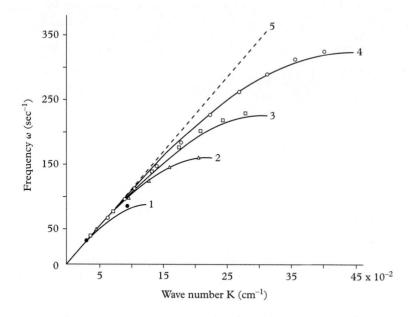

Figure 20.2 *Dispersion relations for different bead separations.*

Higher harmonics in the middle range of the piano keyboard

A convenient set is between key numbers 30 and 50, counting from the left end of the keyboard. Choose one of these keys; a small driving coil placed under one of the relevant three wires excites its modes, while the other two wires are restrained from vibrating by clamping them. A small microphone placed near the string and connected to an oscilloscope picks up the sound and indicates resonance and the exciting of a harmonic when a maximum sound is heard. The frequency is measured accurately and compared with the theoretical formula (20.5). The latter requires measurements and knowledge of the wire parameters entering into (20.5).

Partially unwrapped guitar or piano strings

The frequencies may be measured using a microphone whose output is fed into any spectrum analyzer or data acquisition package. There are many free programs on the internet capable of doing this, for example at <http://download3000.com>, or the Aaronia AG – LCS spectrum analyzer.

Harmonic content in the presence of non-linearities

Between two clamps stretch a metal-wire guitar string whose ends are connected to an oscilloscope. Place a strong magnet across one end of the wire. Pull the wire away from its horizontal position to some distance h, at some distance x_0 from one end, as shown in Fig. 20.3. Place a strong magnet near one end of the wire. When it is released with a suitable mechanism from its initial position, say by burning a string attached to it at x_0, its motion will induce a current which can be amplified and processed for its harmonic content by a digital oscilloscope or other means of spectral (Fourier) analysis. For a fixed x_0, vary h in order to see how the various weights of the harmonics are influenced by the magnitude of h, a parameter that corresponds to increasing non-linearities in the system. These weights, the Fourier amplitudes, may then be compared with the Fourier analysis solutions of the wave equation. Presumably the usual theoretical weights of the harmonics for a given h will correspond better with experimental findings as $h \to 0$. Varying x_0 for a fixed h provides another set of data which may be analyzed in the same spirit.

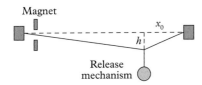

Figure 20.3 *Setup for harmonic analysis.*

REFERENCES

1. F.S. Crawford, *Waves*, Berkeley Physics Course, vol. 3 (McGraw-Hill, New York, 1968).
2. P.M. Morse, *Vibration and Sound* (McGraw-Hill, New York, 1948).

3. P.M. Morse and K.U. Ingard, *Theoretical Acoustics* (Princeton University Press, Princeton, 1987).
4. A.P. French, *Vibration and Waves* (W.W. Norton, New York, 1971).
5. G. Shanker, V.K. Gupta, and N.K. Sharma, Normal modes and dispersion relations in a beaded string, *Am. J. Phys.* **53**, 479 (1985).
6. J.A. Elliott, Intrinsic nonlinear effects in vibrating strings, *Am. J. Phys.* **48**, 478 (1980).
7. J.A. Elliott, Nonlinear resonance in vibrating strings, *Am. J. Phys.* **50**, 1148 (1982).
8. T.D. Rossing, Normal modes of a compound string, *Am. J. Phys.* **43**, 735 (1975).
9. G. Weinreich, Coupled piano strings, *J. Acoust. Soc. Am.* **62**, 1474 (1977).
10. H. Fletcher, Normal vibration frequencies of a stiff piano string, *J. Acoust. Soc. Am.* **36**, 203 (1964).
11. A.H. Benade, *Fundamentals of Musical Acoustics* (Dover, New York, 1990).

21 Physics with loudspeakers

Introduction	124
Theoretical ideas	124
Goals and possibilities	126
Experimental suggestions	127

Introduction

Some hi-fi enthusiasts used to build their own speaker systems, others went into raptures about their quality. Two of a loudspeaker's characteristics may be investigated as a project: its fundamental resonance frequency under different loading conditions, and the angular distribution of the sound field it generates. One may use speakers as microphones, accelerometers, or to study interference and diffraction phenomena, as well as resonant cavities for sound waves. When connected to an oscillator, a loudspeaker is also a convenient source of small-amplitude mechanical vibrations. Some of these uses are set forth below.

Theoretical ideas

For more details on the following, see Refs [1–3].

Resonance frequency

A speaker has a lightweight cone fixed to a metal frame, driven by a coil. The basic structural components are shown in Fig. 21.1.

Assume the coil has length L. The radial magnetic field B of a permanent magnet is perpendicular to the current I through the coil. The force on the cone

Figure 21.1 *Basic speaker components.*

is $F = BIL$. We are dealing here with a combined mechanical and electrical system. The mechanical (velocity) response to F may be formulated by analogy with an RLC circuit [1], through the introduction of a mechanical impedance Z_m, such that the velocity is $v = F/Z_m$ (like $I = V/Z_{el}$), Z_m being the mechanical impedance of cone and coil. Mechanical compliance C_{mech} is the analogue of capacitance, and mass M is the analogue of inductance.

Thus, the fundamental resonance frequency of a speaker f_0 occurs when Z_m is a minimum and, pursuing the electrical analogy, we have

$$f_0 = 1/2\pi (MC_{mech})^{1/2}. \tag{21.1}$$

M will vary with different loading conditions of the speaker, causing changes both in Z_m and in f_0, while the place of resistance in electrical circuits is taken by a radiation resistance R_{rad}.

The coil response will be determined by the two impedances – the electrical Z_{el}, and the mechanical Z_m – because the coil motion in the field B induces a back EMF:

$$BLv = B^2L^2I/Z_m. \tag{21.2}$$

If the applied EMF is ε, the current will be given by the resultant of the two effects [1], so that

$$I = \varepsilon/\left(Z_{el} + B^2L^2/Z_m\right). \tag{21.3}$$

The fundamental resonance frequency of the loudspeaker is its effective low-frequency operating limit, quoted by the manufacturer. Near Z_{el} it does not vary much with frequency f, whereas Z_m does. Hence all one has to do to determine f_0 is to plot the variation of ε/I with f, for at f_0 this quantity will be a maximum, corresponding to a minimum in Z_m. One can study the Q-factor of the speaker resonance around this f_0. Different values for Z_m and f_0 are expected for a speaker in a vacuum, free standing in air, when loaded with small masses attached to the cone, or when mounted in a box.

Field patterns

Radiation field patterns may be studied for a single loudspeaker, and for a bank of four, either connected to a single source for coherent radiation, or to separate independent oscillators for incoherent sources, as described in the experimental section.

All the theoretical expressions for single and multiple diffraction and interference in optics are valid for sound waves as well, but care has to be exercised when applying plane-wave approximations and Fraunhofer conditions in terms of the ratios d/λ, d/L, and L/λ, where d is the source spacing, L the distance from the source, and λ the wavelength of sound. Very small speakers will resemble point sources at large L; their frequency response may restrict measurements to high frequency ranges.

Goals and possibilities

- Measure the fundamental resonance frequency, the resonance curve, and the Q-factor of an 8″ speaker under several loading conditions: when free standing in air; when small masses are loaded symmetrically on the cone; when mounted in an airtight box that can be evacuated to low pressure (3 Torr) by a rotary pump; and when the box is lined with some sound absorbing material, say cotton wool. In the latter case f_0 is lowered (this is one aim of the lining, in addition to absorbing reflections), as the air velocity is lowered by the absorber, tending to make the pressure wave propagation process closer to isothermal than adiabatic.
- Interference: one, two, or more small speakers can be mounted along a line on a hardboard panel large enough to eliminate back reflections, each speaker fitting precisely into a hole in the panel. The latter is mounted on a turntable, enabling one to plot the angular distribution of sound intensity at several meters from it. The sound intensity is recorded by a stationary electret microphone whose output is fed, amplified if necessary, to a recording device. Coherently radiating speakers driven by a single signal generator/amplifier at the same frequency will show interference peaks, whereas speakers driven independently at the same frequency, each by its own generator, will not interfere constructively.
- The sound intensity resulting from two speakers may be measured while the phase of one speaker is shifted with respect to the other by varying amounts between 0 and π (see Fig. 21.2).
- Diffraction from a sound source may be explored in two ways:
  1. Diffraction from a circular "piston," as illustrated in Fig. 21.4. Remove the cone from a single large speaker which is enclosed in a box. Affix one end of a thin lucite rod to the diaphragm and glue a light thin

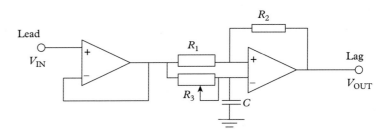

Figure 21.2 *Phase shifter:* $R_1 = 10k = R_2$, $R_3 = 10k$ *pot,* $C = 0.1\,\mu F$.

circular plate to its other end, so that the plate coincides with the plane of the speaker's rim. Using a microphone, plot the direction dependence of the sound radiated by the vibrating plate. The radiation pattern will correspond to that of diffraction at a single aperture.

2. Fresnel diffraction zones for diffraction at a circular aperture may be explored as follows. Terminate a wooden wave box (see experimental section) by a panel with a circular aperture. Explore the output of a microphone as it is moved to and fro along the centerline, the axis of the aperture, outside the wave box. At a minimum, the aperture consists of an even number of zones. Construct masks for various parts of the aperture in order to observe the effects of separate zones.

Experimental suggestions

Basic equipment for all acoustic projects is an audio sine wave generator, an audio power amplifier, an oscilloscope with memory, a small electret or crystal microphone, and loudspeakers of varying diameters. A wave analyzer with FFT and decibel output, or any data acquisition package, will greatly enhance the quality of data analysis.

An anechoic chamber is only a dream for most schools. Nevertheless, experiments in the lab compared to those in the open (for example, the school grounds) will give some indication of the effects of reflection from walls, though not from floors.

For measuring the fundamental resonance frequency of a speaker, only an audio sine wave generator, an AC ammeter and AC voltmeter are needed. The wooden airtight box for a 25 cm diameter speaker may measure $60 \times 35 \times 30$ cm, with a hole for the pump conduit sealed all round.

For interference experiments, a variable spacing arrangement for mounting speakers, capable of being rotated around a vertical axis, is recommended.

For two-speaker interference experiments a simple phase-shift circuit for one of the speakers is shown in Fig. 21.2; a phase shifter can also be bought commercially.

In order to obtain radial plots at a distance of several meters it is useful to have small speakers, say 5 cm diameter, driven at around 10 kHz. Radiation may be detected by a microphone whose signal is fed to a suitable data acquisition system. Mounting the lucite rod and vibrating panel onto a speaker to study diffraction is shown in Fig. 21.3. A speaker of 25 cm diameter, driven at around 3 kHz, is suitable. Here it is preferable to mount the much lighter microphone on the turntable when studying directional effects. Polar paper mounted at the base of the turntable allows one to measure rotation to half-degree accuracy.

The wave box for diffraction at a circular aperture may be constructed from wood as follows: its cross section is 30×30 cm, its length some 2 m, its inside is

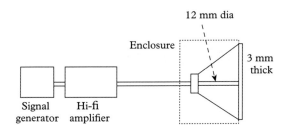

Figure 21.3 *Measuring the diffraction pattern of a circular plate.*

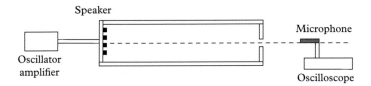

Figure 21.4 *Diffraction and Fresnel zones for diffraction through a circular hole.*

lined with sound absorbing material (rock wool, cotton wool, etc.). One end of the wave box is closed with a panel in which four small speakers are mounted, acting as a coherent source of radiation. The other end of the box terminates in a panel with a 5 cm-diameter hole cut centrally in it. The setup is shown in Fig. 21.4.

....

REFERENCES

1. L.E. Kinsler and A.R. Frey, *Fundamentals of Acoustics* (Wiley, New York, 1962), pp. 247–59, 270–72 (and later editions up to year 2000).
2. L.L. Beranek, *Acoustics* (McGraw-Hill, New York, 1954), pp. 183–90.
3. B.T.G. Tan, Fundamental resonant frequency of a loudspeaker, *Am. J. Phys.* **50**, 348 (1982).

Physics of the tuning fork

22

Introduction

The tuning fork is a rather complex source of sound, comprising two vibrating cantilevers joined at the stem. The starting point for the project is the simple observation that the near-field radiation pattern has four minima and maxima, while the far field has only two. It is then of interest to measure the near and far sound-field intensity distribution for an electrically driven tuning fork and to compare with theoretical models of the radiation. In addition, an electrically driven tuning fork can be forced to vibrate at a range of frequencies around its fundamental resonance, so that it enables one to study experimentally the phenomena of transients, resonance, damping, and beats.

Introduction	129
Theoretical ideas	129
Goals and possibilities	130
Experimental suggestions	131

Theoretical ideas

The angular distribution of loudness (the radiation field) of a tuning fork has been treated theoretically by Sillito [1]. He has shown that the radiation field near the fork, say when it is held near one ear, is determined by path-length-dependent amplitude differences. At distances far from the fork amplitude differences are negligible, and the dominant effect in combining the two waves is the phase difference created by the path difference. The principal mode of a tuning fork, when the prongs vibrate in the plane of the fork, together with regions of loud (A) and quiet (B), are shown in Fig. 22.1.

It will be of interest to compare the experimentally measured angular distribution of the radiation field with theoretical models. In the fundamental mode of the fork each prong may be thought of as an oscillating dipole [3]. The existence of four maxima and four minima is reminiscent of quadrupole radiation. Indeed, the superposition of radiation from two dipoles oscillating out of phase results in a quadrupole. In Crawford's book on waves [3] there is a nice physical discussion of prong movements and the air velocity around them which result in a lateral quadrupole (see Fig. 22.2), but it has been shown experimentally [4] that the angular distribution corresponds best to a linear quadrupole model, described in (22.2), both for the near-field and the far-field pattern.

For a lateral quadrupole consisting of four point sources the sound pressure field $p(r, \theta)$ at distance r and angle θ from the source is given for the fundamental mode by [1,2,5]

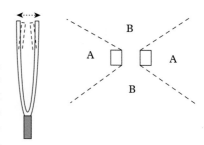

Figure 22.1 *Loud and quiet areas, end view.*

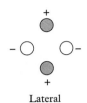

Figure 22.2 *Lateral and linear quadrupoles.*

$$p(r,\theta) = \frac{A}{r}\left(\frac{3}{r^2} - k^2 - i\frac{3k}{r}\right)\sin\theta\cos\theta, \qquad (22.1)$$

where $k^2 = \pi/\lambda$ and A is an amplitude. The imaginary third factor signifies a phase shift of $\pi/2$.

By contrast, the radiation field in the fundamental mode for a linear quadrupole is given by the function [1,2,5]

$$p(r,\theta) = \frac{A}{r}\left[(1 - 3\cos^2\theta)\left(\frac{ik}{r} - \frac{1}{r^2} + \frac{k^2}{3}\right) - \frac{k^2}{3}\right]. \qquad (22.2)$$

These expressions may be analyzed for the near-field $r < \lambda$ and far-field $r > \lambda$ cases, and compared with experimental intensity plots.

Consider now the forced vibration and resonance curve of the tuning fork. In the transient stage the natural frequency of the electrically driven tuning fork will be damped out exponentially, after which it will vibrate at the driving frequency. If the two frequencies are close, beats will be produced, with an amplitude that decreases with time. A frequency sweep toward the natural frequency will result in a sharp increase of the steady state vibration amplitude.

The amplitude curve $A(\omega)$ is a Lorentzian about the resonance frequency ω_0:

$$A(\omega) \propto 1/[(\omega_0^2 - \omega^2) + \omega\omega_0/Q]^{1/2}, \qquad (22.3)$$

where, for Q large enough ($Q > 10$ is a rule of thumb), one can show that the quality factor Q is given by $Q = \omega_0/\Delta\omega$, and $\Delta\omega$ is the width of the frequency interval centered about ω_0. At $\omega = \omega_0 \pm \Delta\omega$, $A(\omega) = A_{max}(\omega_o)/\sqrt{2}$. If the fork is excited by an external force which is then switched off, it will decay according to the formula

$$A(t) = A(t = 0)\exp[-(\omega_0/Q)t]. \qquad (22.4)$$

Goals and possibilities

- As a preliminary experiment we may plot the field for a firmly clamped hacksaw blade (a single dipole), and find the dependence of its frequency on its length.
- Plot by means of a microphone the radial distribution of the pressure field produced by the tuning fork in the near-field, intermediate, and far-field cases in the fundamental symmetric mode. Compare the polar plots thus obtained with quadrupole models of tuning fork radiation fields.
- Repeat for the lower-frequency antisymmetric in-plane mode of the tuning fork, in which the tines move both to the right and to the left at the same time. Try to see whether the resulting pattern may be fitted with a dipole model.

- Study the amplitude resonance curve of the fork by sweeping around its natural frequency with the driver. From the curve derive the quality factor of the resonance.
- Study beats in the forced vibration of the fork for a driving frequency slightly different from the natural frequency, taking particular care to record the passage from the transient to the steady state.
- It is of interest to study the decay of free vibrations; their decay is described by (22.4).
- Additional damping, introduced by placing a small magnet near the stem at the bottom, will affect both the amplitude and the frequency.
- Repeat the above for a tuning fork of much higher natural frequency, say 2000 Hz.

Experimental suggestions

Field pattern of a tuning fork

Obtain as small a 440 Hz tuning fork as possible and mount it on a turntable, clamping its stem firmly. Stick a small neodymium magnet on each of its prongs near the top, to ensure symmetrical loading. Drive the fork with a coil placed opposite one magnet. The coil is connected to a sine wave generator, tuned to the natural frequency of the fork loaded with the magnets, thereby exciting its symmetric mode. A possible setup is shown in Fig. 22.3. Measure the near-field, intermediate, and far-field polar patterns with as small a microphone as possible, which stays in a fixed position, while rotating the fork with the turntable in convenient steps of several degrees, measured by polar paper under the turntable. Near field could be 5 cm, far field about 80 cm (arm's length). Reflections from adjacent surfaces should be minimized: the small dimensions of the fork and the distances involved allow sound insulation to be arranged relatively easily. Insert

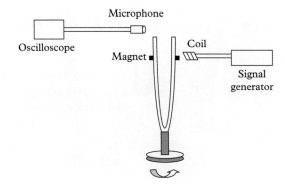

Figure 22.3 *Setup for field pattern.*

baffles of cardboard between the prongs, or cover one prong with a cylinder, in order to see how lobes of minimum are replaced by a maximum as the system is thereby changed from a quadrupole to a dipole.

Frequency response and decay

For measuring the frequency response of a driven tuning fork, use the setup illustrated in Figs 22.4(a) and (b). Neodymium magnets are stuck to the tines, the coils on either side just fit over the magnets. They are made of several hundred turns of enameled 0.15 mm-diameter copper wire, wound on plastic bobbins. One coil, connected to the function generator, drives the fork, the other has induced voltage in it matching the vibrations, which is then analyzed. There is no measurable coupling between the coils. The frequency band over which the resonance curve is measured will be determined by the natural frequency, size, and stiffness of the fork, and the available output from the oscillator, which must be accurate to within 1 Hz or better.

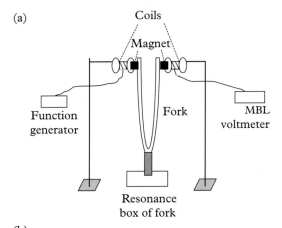

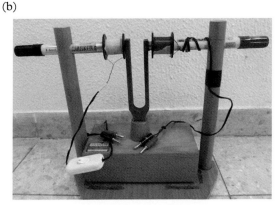

Figure 22.4 *(a) Exciter and detector for frequency response and fork resonance. (b) Photograph of apparatus for resonance experiments.*

REFERENCES

1. R.M. Sillito, Angular distribution of the acoustic radiation from a tuning fork, *Am. J. Phys.* **34**, 639 (1966).
2. A.B. Wood, *A Textbook of Sound* (Macmillan, New York, 1955), p. 66.
3. F.S. Crawford, *Waves*, Berkeley Physics Course, vol. 3 (McGraw-Hill, New York, 1968), p. 532.
4. D.A. Russell, On the sound field radiated by a tuning fork, *Am. J. Phys.* **68**, 1139 (2000).
5. A.D. Pierce, *Acoustics* (Acoustical Society of America, New York, 1989), p. 159.

23 Acoustic resonance in pipes

Introduction	134
Theoretical ideas	134
Goals and possibilities	135
Experimental suggestions	135

Introduction

Acoustic resonance occurs whenever a system (objects excited by wind, rods struck, musical instruments, humans or animals, etc.) produces sound of a particular pitch. Such a sound comprises a fundamental frequency and various admixtures of higher harmonics, these giving it a distinctive quality (the frequency spectrum).

This project investigates the steady state acoustic resonance of air columns in open and closed pipes and in wind instruments. Their response to different frequencies is measured with a microphone and an oscilloscope, in order to explore the effect of various properties of the system on the natural frequencies, and on the sharpness of resonance as quantified by the quality factor Q.

Theoretical ideas

Oscillating resonant systems involve the transformation in each cycle of stored energy from one form to another, accompanied by some degree of dissipation. The latter can take many forms, depending on the system: internal friction due to viscous forces, thermal, radiation, or resistive losses.

The normal mode (circular) frequencies ω_{0n} for cavities and pipes are determined by their shape and size (boundary conditions), as well by the way in which the exciting mechanism, the driver, is coupled to the system.

The quality factor is formally defined as the ratio of energy stored to that lost per cycle. The smaller these losses, the larger Q. It is therefore a measure of the sharpness of resonance. Basic formulas are discussed in the project for the tuning fork.

Theoretical expressions for Q in acoustic systems relevant for the present project are derived by exploiting the extremely fruitful analogy between the electrical and acoustic parameters (acoustic resistance and impedance, pressure amplification), for which the reader is referred to the literature [1].

Finally, we remark that Q also governs the rate at which transients build up to the steady state in a driven system, or the rate of free decay from the steady state. It would be a useful part of the project to try and fit the data with the resonance Lorentzian line shape and to discuss any reasons for departures from it.

Goals and possibilities

- Investigate resonance frequencies for the fundamental and for harmonics in pipes of the same lengths but differing diameters (end effects are diameter dependent).
- Scan the signal amplitude from a small microphone connected to an oscilloscope around each resonance frequency, and establish the resonance curves.
- Calculate Q for various pipes and wind instruments for different modes of excitation (closed and open pipes, with and without mutes).
- Measure response as a function of length.
- Attach a flange to one end of the pipe to see its effect on end corrections.
- Try pipes of different materials (cardboard, glass, plastics, metal) in order to see effects of wall rigidity and surface roughness on Q [2].
- Vary the coupling of the speaker to the pipe to see its effect on Q (see the experimental suggestions).
- Construct a model whose parameters, fitted from experimental data, could guide the behavior for higher harmonics.
- Study the matching effects of horns.

Experimental suggestions

The basic tools are an audio signal generator capable of driving a loudspeaker (or one in tandem with an audio amplifier), a digital frequency meter, a loudspeaker, two small microphones around 10 mm in diameter so as not to disturb the sound field, and an oscilloscope or data acquisition package.

To ensure reproducible uniform energy input, channel the sound from the speaker by means of a funnel of suitable diameter (the top end of a soft-drink bottle can do the trick) through a hole in a rigid plate covering one end of the pipe. Position one microphone at the input to monitor the energy input from the speaker (which, together with the signal generator output, may be frequency dependent). The size of this hole may be varied, to see what influence this coupling has on Q [2].

Sound at the far end is detected by the second microphone, with or without the use of a similar cover plate, depending on the mode of excitation. The microphone can also be moved around to explore the inside of the resonator. Figure 23.1 illustrates a possible experimental setup.

Figure 23.2, from [2], shows the resonance curves for pipes of the same dimensions, one plastic, the other cardboard. Note the different Q values, 38 and 31, respectively.

136 *Physics Project Lab*

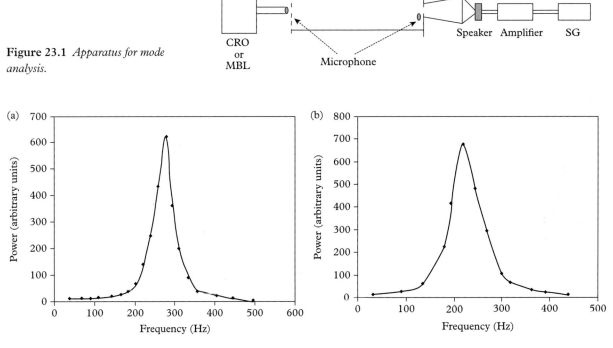

Figure 23.1 *Apparatus for mode analysis.*

Figure 23.2 *Resonance curves in pipes: (a) plastic, (b) cardboard.*

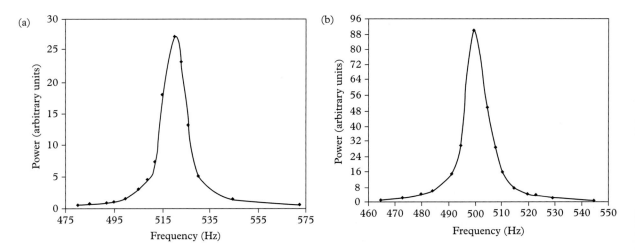

Figure 23.3 *Resonance curves: (a) open end, (b) partially closed end.*

Figure 23.3 shows the effect of partially closing the end of the same pipe, resulting in a change of losses and therefore of Q. For the change from open to partially closed pipe the change in Q is from 38 to 51, and in the mode frequency from 520 Hz to 500 Hz.

....

REFERENCES

1. L.E. Kinsler and A.R. Frey, *Fundamentals of Acoustics* (Wiley, New York, 1962), and later editions up to year 2000.
2. P. Gluck, S. Ben-Sultan, and T. Dinor, Resonance in flasks and pipes, *Phys. Teach*. **44**, 10 (2006).

24 Acoustic cavity resonators and filters

Introduction	138
Theoretical ideas	138
Goals and Possibilities	139
Experimental suggestions	140

Introduction

This project deals with acoustic filters and the modes of vibration of Helmholtz resonators. It has technological relevance to silencers (mufflers), and to attenuation and transmission coefficients in pipes.

Theoretical ideas

Cavity resonators

The frequency of the fundamental mode of a Helmholtz resonator, consisting of a cavity of volume V ending in a narrow neck of effective length L and cross-sectional area S, is given by [1]

$$f = (c/2\pi)\sqrt{(S/LV)}, \tag{24.1}$$

where c is the sound velocity. A useful alternative form is

$$kL = \sqrt{(v/V)}, \tag{24.2}$$

where v is the volume of the neck (depends slightly on relative humidity) and k is the wavenumber. Note that f is independent of the shape of V; for a given opening only the size of V is relevant. This mode can be understood in terms of the oscillations of the mass of gas in the neck. Higher modes, overtones, can result from standing waves in the cavity, and are therefore shape dependent. In addition, other frequencies may be observed if the neck is long and one changes the ratio v/V by gradually filling the main volume with water. Such cases have been treated by means of models of increasing complexity in a beautiful article by Crawford [2], and are given by the solutions of

$$kL \tan kL = v/V. \tag{24.3}$$

The quality factor Q of a driven Helmholtz resonator is given by [1]

$$Q = 2\pi\sqrt{(L^3 V/S^3)}. \tag{24.4}$$

Acoustic filters

The basic idea of filters is to change the acoustic impedance in a pipe or duct when it forms a junction with a pipe of different diameter, or with a Helmholtz resonator, or if there is an opening in the pipe somewhere along the line, as in Fig. 24.1. These cause reflections, storage, and dissipation of energy, and a reduction of energy transmitted beyond the junction. For example, if a Helmholtz resonator of frequency f is coupled into a pipe, there is a drastic dip in the transmission coefficient of the system at the resonant frequency f. By appropriately choosing orifices and resonant volumes, low- and high-pass filters may be built, an experimental goal of the project.

Figure 24.1 *Filter model.*

Goals and Possibilities

- Explore the validity of (24.1) for various bottles or flasks whose necks are uniform narrow cylinders joining the main volume abruptly, for example volumetric flasks. A simple way of confirming (24.1) is described in the section on experimental suggestions.
- For a spherical 500 cc volumetric flask with a long (15–20 cm) neck, measure, using a microphone, the frequencies of the lowest and of higher modes as the flask is filled with water (see below) when the flask is excited by tapping, and compare with (24.4).
- Drive a Helmholtz resonator near the lowest mode, measure the response with a microphone, in order to find Q, and compare with (24.2).
- Measure the transmission coefficient of energy (proportional to the pressure amplitude squared) down a 3 cm-diameter tube by recording the ingoing and outgoing signals with two identical microphones connected to an oscilloscope, when the tube is excited by a loudspeaker near one of its resonant modes. Now couple a 100 cc spherical flask with a 6 mm-long and 1.5 cm-diameter straight neck into a hole drilled into the tube, as shown in Fig. 24.2, and measure the transmission coefficient as a function of f around the resonant frequency of the flask, in order to observe the sharp dip in transmission – the acoustic filter effect.
- See how, when the flask in the hole is replaced by a tube of the same dimensions as the flask neck in the previous experiment, the tube is converted into a high-pass filter.
- Try interposing a short tube of half the diameter between two sections of a 3 cm-diameter tube, as shown in Fig. 24.3, and explore the response of the system over a range of frequencies.

Figure 24.2 *Acoustic filter with Helmholtz resonator.*

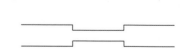

Figure 24.3 *Effect of constriction.*

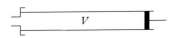

Figure 24.4 *Speaker, flask, and microphone.*

Figure 24.5 *Measuring the fundamental mode of a Helmholtz resonator.*

Figure 24.6 *Tapping a flask with a mechanical vibrator – the microphone is on top, for mode analysis.*

Experimental suggestions

Befriend a glass blower. He will have a set of neckless spherical flasks, like the one shown in Fig. 24.4, onto which he can join necks of varying lengths and diameters. Alternatively, chemical supply companies have standard stocks of many sizes for v and V (Florence flasks). Collect wine, beer, or liqueur bottles with sharp demarcations of uniform necks from the main volume.

Make sure that the wavelengths of sound in this project are always much larger than typical apparatus dimensions. This enables comparison with theoretical calculations that are valid for this simpler case.

Equation (24.1) can easily be tested by constructing a lid with a short narrow pipe to fit onto a tube equipped with a plunger, as in Fig. 24.5. Using a set of tuning forks or a loudspeaker, and moving the plunger, excite the system to its resonant frequencies. Measure the volumes and plot V as function of $1/f^2$. The result should be a straight line not going through the origin, indicating the presence of an effective volume comprising also that of the neck.

Since tapping the flasks causes freely decaying oscillations lasting a second or less, a way has to be devised [3], shown in Fig. 24.6, for recording the sound with the microphone joined to the wave analyzer at the same time. This should be worked out by some triggering mechanism. In order to be able to perform an FFT spectrum analysis on the signal, clearly a high sampling rate is required. These are available on most data acquisition systems.

REFERENCES

1. L.E. Kinsler and A.R. Frey, *Fundamentals of Acoustics* (Wiley, New York, 1962), and later editions up to year 2000.
2. F.S. Crawford, Lowest modes of a bottle, *Am. J. Phys.* **56**, 702 (1988).
3. P. Gluck, S. Ben-Sultan, and T. Dinor, Resonance in flasks and pipes, *Phys. Teach.* **44**, 10 (2006).

Room acoustics

25

Introduction

Any enclosure in which sound is generated and is reflected from the walls is a system in which resonance and standing waves can be set up, damping of sound waves is present, and reverberation time is of interest. This project simulates such a system by exploring the resonance properties and associated phenomena of a rectangular closed box.

Introduction	141
Theoretical ideas	141
Goals and possibilities	142
Experimental suggestions	143

Theoretical ideas

The wave equation for the excess pressure $p(\vec{r}, t)$ is given, in the absence of damping, by

$$\nabla^2 p(\vec{r}, t) + \frac{1}{c_s} \frac{\partial^2 p}{\partial t^2} = 0, \tag{25.1}$$

where c_s is the velocity of sound. The normal modes are determined by the boundary conditions. In a rectangular cavity whose walls are at $x(0,a)$, $y(0,b)$, $z(0,c)$, the appropriate solution is

$$p(x, y, z, t) = A e^{i\omega t} \cos \frac{l\pi x}{a} \cos \frac{m\pi y}{b} \cos \frac{n\pi z}{c}, \tag{25.2}$$

where the normal mode frequencies f are given by the triad of natural numbers l, m, n:

$$f_{lmn}^2 = (c_s^2/4)\{(l/a)^2 + (m/b)^2 + (n/c)^2\}. \tag{25.3}$$

Unlike for a one-dimensional string, the frequencies are not multiples of a fundamental.

The number of normal mode resonances $N(f)$ in the frequency range 0 to f is given by the following approximate expression in the book by Morse and Ingard [1]:

$$N(f) = (4\pi V/3c_s^3)f^3 + (\pi A/4c_s^2)f^2 + (L/8c_s)f, \tag{25.4}$$

where V is the volume of the chamber, A is the total surface area of its interior, and L equals $4(a+b+c)$. Typically, for frequencies above 1 kHz, $N(f)$ increases rapidly, and the individual resonances are partially smeared out and start to blend into a general background level in the chamber (this would depend on the size of the chamber relative to the wavelength).

When damping is present, the equation is modified. If one assumes a model in which dissipation is proportional to dp/dt, separation of variables leaves the spatial part of the wave equation unaffected, whereas the temporal part $T(t)$ satisfies

$$\frac{d^2 T}{dt^2} + \Gamma \frac{dp}{dt} + \omega_0^2 T = 0, \tag{25.5}$$

where $\omega_0 = 2\pi f_{lmn}$, and Γ is a damping constant. Standard analysis for the case $\omega_0^2 - (\Gamma/2)^2 > 0$ gives the solution

$$T(t) = B \exp\{-\Gamma t/2\} \sin\{[\omega_0^2 - (\Gamma/2)^2]^{1/2} t + \varphi\}. \tag{25.6}$$

Here, B and φ are fixed by initial conditions. Thus, free oscillations decay with a time constant of $2/\Gamma$.

If a forcing term $F\cos(\omega t)$ is added to the right-hand side of (25.5), one encounters the usual resonance phenomena, the solution now becomes $B\cos(\omega t + \alpha)$, and the amplitude B satisfies

$$B^2 = \frac{F^2}{\omega^2 \Gamma^2 + (\omega_0^2 - \omega^2)^2}. \tag{25.7}$$

A similar expression was mentioned in the project on tuning forks and is discussed in detail in the book by French [4], where the parameter Γ is shown to be related to the quality factor Q, and to the frequency width $\Delta\omega$ at which power is half the maximum value delivered to the system by the external source.

Goals and possibilities

The general aim is to understand as completely as possible the normal mode excitations, their free decay, the transients, and the resonance properties of a rectangular enclosure.

- Measure as many normal mode frequencies as possible and compare with the theoretical prediction, (25.3). In this way, determine the velocity of sound and compare with the empirical relation $331 + 0.6t$, where t is the temperature in Celsius.
- How many resonances N can one detect in the frequency range 0 to f, and how does this compare with the theoretical value for $N(f)$ in (25.4)?
- Map nodal planes in the enclosure for a particular frequency or triad (l,m,n) and compare with the theoretical result (25.2).

- Excite some modes and measure the decay times and decay envelopes as functions of the mode frequencies.
- Look at beat phenomena when the external driver frequency differs from the natural one.
- Establish resonance curves around several normal modes of the enclosure and calculate the Q values.
- Construct an identical enclosure in which the walls are less smooth, and look at the resulting losses and changes compared to the previous results.
- Insert into the chamber rectangular objects whose dimensions are small compared to the chamber, in order to see how resonances are perturbed in frequency. Objects located near the center reduce f, while those near the ends of the chamber increase it [3].

Experimental suggestions

The basic equipment comprises the following items: a small lapel microphone 1 cm in diameter, a good quality 3″ loudspeaker, a function generator capable of driving it in the range 0.2–2.5 kHz with a resolution better than 1 Hz, and a data acquisition system with FFT capability.

A convenient rectangular box, sketched in Fig. 25.1, is constructed from a wooden top and bottom and at least two glass-pane sides, enabling one to observe the position of the microphone probe as one maps the modes. A suitable size would be 65 × 50 × 25 cm. One side could be fitted with hinges in order to be able to open and reseal the box, enabling one to change the texture of the walls by lining with different materials, and thereby explore the resulting effect on mode damping. The box is placed on a platform supported by rubber to minimize vibration. Weights may be placed on top of the chamber to optimize the amplitude of resonances. The speaker can be coupled to the box in several ways. The simplest, though not recommended for a large speaker, is to insert it straight into a hole of suitable diameter in one face. Instead, insert a funnel outlet into a small hole in the wall, with the speaker attached to the other end, as shown in Fig. 25.2. This makes it more like a point source. Two such holes are made in the box, say at a corner and a face center, each covered with a suitable flap when not in use. You can then drive the interior at different places to see what effect this has. The back of the speaker may be enclosed in damping material.

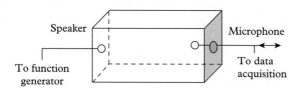

Figure 25.1 *Setup for mode mapping.*

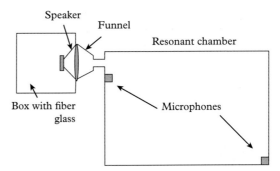

Figure 25.2 *Coupling the speaker to the box.*

The microphone is attached to the end of a thin metal rod or fitted at the end of a thin tube of the same inner diameter as the outer diameter of the microphone. This probe is inserted through a hole in the wall opposite the speaker, so as to be able to reach most points in the interior. Connecting wires are fed through the thin tube or attached to the rod. The hole in the middle of one wall is covered with felt or rubber through which the tube is manipulated.

The probe is first held diagonally across from the speaker and the lowest mode at which the box resonates is found, the (1,0,0) mode, the long dimension being the x direction. The speaker amplitude is kept constant, and the frequency is varied around resonance to obtain a typical resonance curve. Moving the probe around enables location of the nodal planes and so indicate which mode has been excited. You can then move to higher resonance frequencies. Mode coupling of two close modes may occur, meaning interaction between modes and exchange of energy between them. Nodal planes would then be be difficult to locate, and in this case one should move to another mode.

It is a useful facility to be able to drive the speaker with a function generator which has a voltage-controlled frequency sweep, by means of a ramp generator circuit, at the same time monitoring the frequency. One could then feed the microphone output to a digital scope and display the whole resonance spectrum. A count of the resonance peaks in a certain frequency range $N(0, f)$ enables one to compare it with the theoretical expression, (25.4).

If the speaker is driven by a square wave of variable frequency, resonance peaks will be observed when the square wave frequency f_{sq} satisfies $f_{sq} = f/n$ and n is an odd positive integer. We shall thus have harmonics of the resonant modes of the chamber, a neat way of seeing Fourier decomposition.

Transients and beats are readily observed when the excitation frequency is close to a natural mode. If the forcing speaker is turned off, one can observe the decay, and calculate Γ and Q. One may try to relate the decay time of free excitations to the reverberation time τ of the enclosure, defined in architectural acoustics by Sabine as the time for the amplitude of the sound to decrease to one thousandth (and the power to one millionth) of the original value: $\tau = (2/\Gamma)\ln(1000) = 13.8/\Gamma$.

For geometrically and acoustically similar enclosures, if the box is a scaled-down version of a room, the reverberation time is expected to scale with dimension. One could then try to predict τ for a room.

REFERENCES

1. P.M. Morse and K.U. Ingard, *Theoretical Acoustics* (McGraw-Hill, New York, 1968), pp. 585–7.
2. A.P. French, *Vibrations and Waves* (W.W. Norton, New York, 1971).
3. P.M. Morse and H. Feschbach, *Methods of Theoretical Physics, Part II* (McGraw-Hill, New York, 1953), pp. 1038–64, 1166.

26 Musical instruments: the violin

Introduction 146
Theoretical ideas 146
Goals and possibilities 146
Further explorations 147
Experimental suggestions 148

Introduction

Exploring the acoustic properties of musical instruments will appeal to a wide circle of researchers. Such a project would provide an opportunity to gain first-hand experience with and deepen understanding of many fields of physics which are only cursorily touched upon in introductory or even physics major courses: wave amplification by resonant cavities, decay times, spectral analysis, elasticity and stiffness properties of plates, and directional dependence of acoustic radiation. Here we give an example of what can be explored for a violin, but with suitable modification these ideas could be applied, using the same basic experimental techniques, to other string instruments (see the next chapter on the guitar).

Theoretical ideas

Before embarking on the project, or as it develops, some understanding of normal modes of vibration of strings and bars, membranes, and cavities should be nurtured by the instructor. Ideally this should include Fourier analysis, radiation patterns, quality factor, and resonators [1,2]. We refrain from including the relevant theory here – part of it has been given in the project on wires and strings earlier in Part 3, and in any case is available in the references cited. There are many web sites giving a tutorial introduction to the physics of the violin.

Goals and possibilities

- Measure the tension, mass per unit length, and length of any string, in order to compare theoretical and calculated vibration frequencies.
- Pluck the string at different places in order to see which harmonics are present as a function of plucking position. Try $L/2$, $L/4$, $L/5$, and so on, where L is the string length.
- On the violin pluck and bow at the same place, in order to study the difference in "color" of the notes, the harmonic content, generated in different ways.

- Study the different motions of the violin string when bowed, by watching the motion of a suspended second bow resting lightly on the string at three different places: at the center, near the bridge, and near the nut (the "follow-bow" experiment). One can also examine the string with a low-power traveling microscope with stroboscopic illumination.
- A sound-level meter positioned at a fixed distance from the violin can be used to explore the sound intensity as a function of frequency when plucking successive notes on the violin with as equal initial displacements as possible. Does the same note played on different strings give the same intensity?
- A string can be driven at various frequencies either with a coil, if it's magnetic, or if catgut by winding a small piece of iron wire around it and glueing in place. The polar radiation pattern and the resulting sound spectrum can be observed at various locations in the room, ideally an anechoic chamber. A box lined with sound absorbent might work.
- The back and sides of a horizontally held violin may be embedded in a sandbox, the f-hole closed, then excited. This will enable one to isolate the sound spectrum of the top plate.
- Generate sound opposite the f-hole, insert a microphone into the body and discover its Helmholtz resonance. This can be compared to theory by estimating the volume of the sound box and measuring the geometry of the holes.
- A thin wooden violin-shaped top plate can be prepared and its Chladni-type excitations studied. Varnish applied to the plate will affect the stiffness constants, change the sound spectrum, and may give some idea of the importance of varnishing in real instruments.

Further explorations

The above list by no means exhausts the possibilities of exploring the properties of such a complex instrument comprising many parts. Here are some more topics, each of which may be pursued at various levels.

- How does the material of the bow (the horsehair), its tension, and the kind of rosin applied to it affect the sound?
- Can a violin maker provide the experimenter with a realistic top or bottom plate, which can be suitably clamped and then excited by a loudspeaker or mechanical vibrator, in order to study its modes?
- Can bridges be made of different types of wood (possessing different elastic stiffness constants) in order to explore their effect on string-to-body coupling?

- Air and helium mixture can fill the resonant cavity in order to see its effect on the Helmholtz resonance.
- Record and analyze the spectrum of harmonics in a vibrato note (rocking the finger stopping the string backward and forward): the pitch varies up and down, together with the harmonics – this is one of the most characteristic features of the violin sound.
- Reduce the effectiveness of the bridge in power transmission by attaching a mass (called the mute) to it, making the violin quieter and less bright, and analyze the resulting spectrum and intensity.

Experimental suggestions

A good quality signal generator able to drive a loudspeaker (with an amplifier if necessary), a range of small microphones and loudspeakers having suitable responses (measuring and calibrating these is itself a project), a data acquisition package with FFT facility, and a sound level meter are the basic tools for projects of this type. Sophistication can be increased depending on budget and expertise, but we do not expect holographic interferometry or anechoic chambers to be available to most researchers.

The tension in the string can be measured by suspending a small weight from the center of a string and measuring the deflection with a millimeter scale, a micrometer, or a sonic ranger. The coils to excite the strings and detect the vibrations by induction can be home made, or they are also available from supply houses.

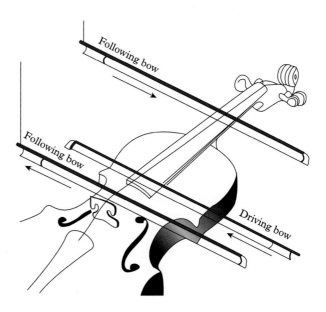

Figure 26.1 *Follow-bow experiment.*

The harmonic content of plucked and bowed strings may be analyzed either by recording on tape and feeding the recording into a wave analyzer, or recording by a microphone whose output is fed into a data acquisition package.

The follow-bow experiment [3] is illustrated schematically in Fig. 26.1 [4]. It can be video recorded for detailed study.

The violin is supported horizontally. The second bow, which serves as the probe, is suspended vertically at its grasping end, while its far end rests very lightly on the string to be bowed. Now bow the string loudly with the driver bow. It is found that the suspended bow moves in the direction of the string's motion during the longer part of each cycle. When placed near the bridge the hanging bow moves in unison with the driver bow, but when placed close to the nut it moves in antiphase. When put at the center of the string there is almost no motion. This kind of behavior is related to the velocity versus time graphs of a bowed string at the three positions, as sketched in Fig. 26.2. Thus the suspended bow follows the slow motion of the string in one direction, but not the fast return of the string in the other direction, so that in each cycle it will suffer a net displacement.

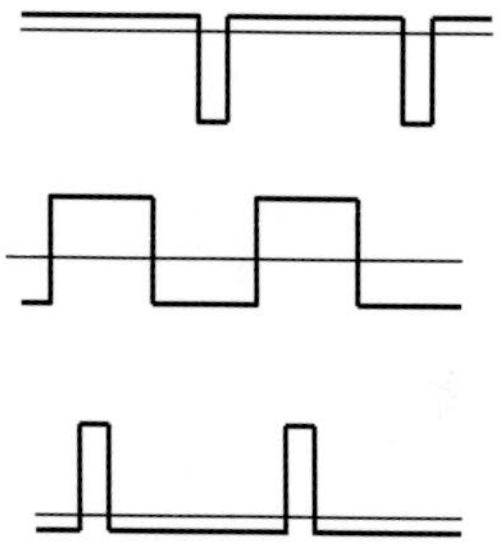

As in the chapter "Vibrating wires and strings," one may excite one of the violin strings electromagnetically to increasing amplitudes. At large amplitudes the stiffness of the string causes non-linearities, whereby the frequencies of the harmonics will not be multiples of the fundamental. These harmonics may be obtained by attaching a phonograph pickup (yes, there are still many old phonographs lying around – grab one) to the back plate of the violin and analyzing the resulting signal. It is then of interest to obtain the frequency deviations Δf from a perfect string as a function of the harmonic number n. Theoretically [5], Δf is proportional to n^3. Figure 26.3 shows the result of such an experiment on the D string. The data are best fitted with a cubic polynomial.

Figure 26.2 *Velocity–time graphs at various positions.*

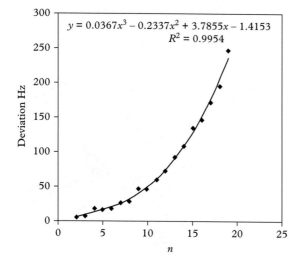

Figure 26.3 *Deviation due to non-linearities as a function the harmonic number n.*

REFERENCES

1. A.P. French, *Vibrations and Waves* (W.W. Norton, New York, 1971).
2. L.E. Kinsler and A.R. Frey, *Fundamentals of Acoustics* (Wiley, New York, 1962), and various editions up to year 2000.
3. J.C. Schelleng, The Physics of the Bowed String, *Sci. Am.* **230**, 87 (1974).
4. T.D. Rossing, F.R. Moore, and P. Wheeler, *The Science of Sound* (Addison-Wesley, San Francisco, 2002).
5. University of New South Wales, Music science: research projects and scholarships, online at <http://www.phys.unsw.edu.au/music/research.html>.

Musical instruments: the guitar

27

Introduction	151
Theoretical ideas	151
Goals and possibilities	152
Experimental suggestions	153

Introduction

Much of what we suggested in the project for violin acoustics is relevant to the guitar. Nevertheless, there are significant differences in their constructions and response, in their popularity, in the amount of experimental and theoretical research literature, and in their mode of playing, which warrants discussion of an independent guitar project.

Theoretical ideas

The coupling between various components of the guitar produce radiated sound in a range of frequencies. Extensive research in guitar acoustics [1–4] has shown that at high frequencies it is the top plate coupled to the strings by the bridge which radiates most efficiently. The low frequency spectrum is mainly the result of radiation by the air cavity and sound hole and that of the back plate. These findings should be kept in mind if one of the goals of the project were to study the roles played by various single components of the guitar and their frequency response.

Since the dominant mode of playing is by plucking the strings, more attention should here be paid to the propagation of pulses in plucked strings than for violins (pizzicato). This was first elucidated by Helmholtz. He showed that a kink propagates back and forth along the string from nut to bridge, along a parallelogram boundary, the result of two pulses traveling in opposite directions. The harmonic components of the kink depend on the place of plucking along the string. One may pluck the string parallel with or perpendicular to the bridge (any other direction being a superposition of these). These two "polarizations" will elicit very different responses, since the top plate responds more readily in the perpendicular plane. An adequate discussion is readily accessible in Refs [4,5]. Plucking may be done with the finger or with metal or plastic plectrums (picks), and these different techniques will affect the sound spectrum.

During the vibration of a string at normal plucking amplitudes, the dominant force exerted by the string on the bridge is expected to be mostly transverse, but if the guitar is played very loudly the changes in the length of the string may in addition cause significant longitudinal forces [6,7].

The project could include mathematical modeling of the guitar by coupling strings, air volumes, and lumped masses, in order to try and reproduce some of the spectral characteristics of its acoustic radiation. For guitars with metal strings the usual assumption of perfect flexibility is not valid and their stiffness would have to be taken into account in any modeling. It is then strongly suggested that previous efforts in this field should be consulted [8–11].

Goals and possibilities

Refer to the violin project for some of the experiments that are relevant here too.

- Strings by themselves radiate very little. Their response when isolated from their complex interaction with the instrument can be measured in the usual way by a sonometer setup.
- Pluck the strings parallel with and perpendicular to the bridge in order to compare the resulting intensity at a fixed distance and direction from the guitar.
- Plucking is a complicated mode of excitation: it may be done with the finger, fingerpicks, or with metal or plastic plectrums (picks), and these different techniques will affect the sound spectrum. All these could be investigated.

As an alternative to analyzing the response of the top plate, apply a controlled sinusoidal force with a suitable force transducer to the top plate on one side of the bridge, at various frequencies. Then explore the following phenomena, preferably in an *anechoic* enclosure, or at least in a reasonably quiet and echo-free space.

- Measure opposite the sound hole, in one particular direction and at a particular distance how, in response to the transducer, the guitar radiates sound waves as a function of frequency. One may thus discover some of the low frequency resonances of the guitar.
- Measure the angular variation of the sound level for any of these frequencies. Different frequencies will result in dipolar, quadrupolar, or some other intensity plots (in an ordinary room this will not work, because of multiple reflections).
- Modify the size of the sound hole so as to be able to measure the lowest Helmholtz frequency (a function of the hole area and the enclosed volume) of the enclosure as a function of hole area.
- Build a simple box model guitar with a single string. Measure the role of the enclosed air coupling to the guitar body in the overall sound spectrum in two cases: (a) when the rib (the piece of wood between the body and

the bottom) is airtight, and (b) when the rib is thoroughly perforated with drilled holes so as to minimize the air coupling to the body.

- Modify the mounting of the string in the box model so as to incorporate a force probe, in order to gather some information on the static and dynamic forces exerted on the bridge. Static forces exerted by the strings on the bridge are over 100 N for steel and 50 N for nylon strings. Measuring dynamic forces as functions of time when the string is plucked in various ways would be one of the highlights of the project.

- Explore the air resonances of the cavity.

- Unlike for the violin, many types of guitar are played: classical and flamenco having nylon strings, folk and archtop equipped with steel strings (these can be driven using a small permanent magnet near the string through which a sinusoidal current is passed). The electric guitar with its electromagnetic pickup offers further possibilities. Comparison between any two types of any of the guitar characteristics could be an interesting part of the project.

Experimental suggestions

A digital sound level meter is useful for measuring sound levels. Alternatively, a small microphone connected to a data acquisition system at a suitable sampling rate will serve the same purpose.

An audio power amplifier driven by an audio signal generator and capable of 20 W output into 8 Ω can establish currents of order 1 A (a resistor in series with the wire, or a 12 V car brake light, will provide a suitable load for the amplifier). A small magnet with poles 10 mm in diameter and field 0.1 T can produce a transverse sinusoidal force of 10^{-3} N to drive the string in different modes. Amplitudes can be measured with a micrometer (advanced till it just touches the wire), or a magnifier with stroboscopic illumination.

A motor, geared down to one revolution per second and firmly held in an adjustable mounting above the string, could rotate an arm with a plectrum so as to excite the string in a reproducible manner.

Embed the guitar in a sand box and place weights on the top plate to freeze movements while exploring air resonances. Join a flexible plastic tube suitably coupled (part of the top of a soft-drink plastic bottle is a suitable funnel) to a loudspeaker connected to a signal generator/amplifier. The tube can be inserted at various places into the enclosure so as to generate pressure variations at the chosen frequency. Small electret microphones suitably placed inside and connected to a data acquisition system can then map those pressure variations and locate air resonance frequencies.

When building a model box guitar some basic data for all guitars should be remembered. The top plate is usually 2.5 mm thick, often made from spruce. The back is the same thickness, made of some hardwood, like maple. Strings are usually 65 cm long. Various bracings are used for both plates in order to strengthen them. Expert advice should be sought on building an anechoic space from scratch.

....

REFERENCES

Note: Guitar acoustics research is an active field. The many web sites should be consulted for more details and research possibilities.

1. T.D. Rossing and G. Eban, Normal modes of a radially braced guitar determined by TV holography, *J. Ac. Soc. Am.* **106**, 2991 (1999).
2. T.D. Rossing, J. Popp, and D. Polstein, Acoustical response of guitars, in *Proceedings of the Stockholm Musical Acoustics Conference 1983*, pp. 311–32 (1985).
3. B.E. Richardson and G.W. Roberts, The adjustment of mode frequencies in guitars: A study by means of holographic interferometry and finite element analysis, in *Proceedings of the Stockholm Musical Acoustics Conference 1983*, pp. 285–302 (1985).
4. T.D. Rossing, F.R. Moore, and P.A. Wheeler, *The Science of Sound*, 3rd ed. (Addison-Wesley, New York, 2002).
5. J.C. Schelleng, The physics of the bowed string, *Sci. Am.* **230**, 87 (1974).
6. E. Watson, *Transverse and Longitudinal Forces of a String on the Guitar Bridge* (Northern Illinois University Press, DeKalb, IL, 1983).
7. N.H. Fletcher and T.D. Rossing, *The Physics of Musical Instruments*, 2nd ed. (Springer, New York, 1998), ch. 9.
8. G. Caldersmith, Guitar as a reflex enclosure, *J. Ac. Soc. Am.* **63**, 1566 (1978).
9. K.A. Stetson, On modal coupling of string instrument bodies, *J. Guitar Acoustics* **3**, 23 (1981).
10. O. Christensen, Quantitative models for low-frequency guitar function, *J. Guitar Acoustics* **6**, 10 (1982).
11. I.M. Firth, Physics of the guitar at the Helmholtz and first top-plate resonances, *J. Ac. Soc. Am.* **61**, 588 (1977).

Brass musical instruments

28

Introduction

Exploring the physics of the trumpet or trombone is an example of what can also be done with other members of the brass family of instruments like the tuba, French horn, and so on. The motivation for and success of such a project would be enhanced by the participation of an active player of the relevant instrument who would be able to pay proper "lip service" to the sounds produced during experimentation. A fair knowledge of musical scales and the ability to recognize various musical intervals would also be an advantage.

Introduction	155
Theoretical ideas	155
Goals and possibilities	157
Experimental suggestions	159

Theoretical ideas

We refer to several widely available books where the interested reader can find thorough discussions of both theory and experiments on the acoustics of musical instruments. These are the classics of Helmholtz [1], Benade [2,3], and Backus [4], and the more recent texts of Rossing and co-workers [5,6]. Our discussions will therefore be restricted to the bare minimum.

The player's lips perform the function of a vibrating reed. When air is blown into a brass instrument while the player's lips are vibrating at or close to one of the instrument's resonance frequencies, a standing wave is set up in the air column. A feedback mechanism comes into play: the excited air column drives the lips to maintain the vibrations. A complex tone results, with a fundamental frequency which is close to one of the instrument's resonance modes. Higher resonances of the air column result from increased lip tension. The playing frequencies are close to the actual resonances.

Brass instruments are mostly cylindrical tubes, but the cup-shaped mouthpiece and the flared bell induce marked differences from what one would expect from the modes of a closed pipe.

Consider first the mouthpiece. When closed by the lips it behaves like a Helmholtz resonator: the mouthpiece cup being the cavity of volume V, the tapered tube joined to it serving as the constriction of cross-sectional area S and effective length L_e. The resonance frequency of such a resonator is (see Chapter 24)

$$f = \frac{1}{2\pi}\sqrt{\frac{c^2 S}{L_e V}}, \qquad (28.1)$$

where c is the speed of sound, $S = \pi r^2$, and L_e is the sum of the actual length L of the tapered tube and the end correction $0.8r$, r being the mean tube radius. This frequency may be heard by slapping the mouthpiece on the palm of the hand (hence the nickname "popping frequency").

The derivation of the effective length added by the mouthpiece to the tube to which it is attached is rather lengthy, and we refer you to Refs [7,8]. It is given by the expression

$$L_f = \frac{c}{2\pi} \tan^{-1}\left[\frac{2\pi f L_0/c}{1 - (4fL_1/c)^2}\right], \quad (28.2)$$

and is frequency dependent. In this expression L_0 is the length of a cylindrical tube of the same cross-sectional area as the tubing to which the mouthpiece is attached that has the same volume as the entire mouthpiece. It is the minimum effective length that the mouthpiece adds to the air column. L_1 is the length of the same cylindrical tube closed at one end that has the same fundamental frequency as the popping frequency of the mouthpiece. It is the maximum effective length of the mouthpiece. One may calculate L_f for both trumpet and trombone,

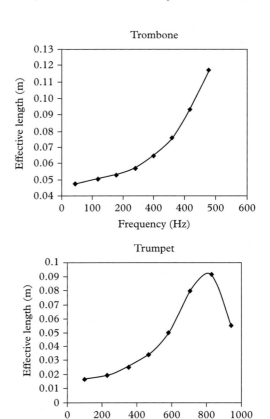

Figure 28.1 *Frequency dependence of effective lengths.*

and Ref. [8] shows how to do this. The result is an increase in effective length with f, which means that the mouthpiece affects (lowers the frequencies of) the higher modes more than the lower ones. Figure 28.1, reproduced from [8], shows plots of effective lengths of a trombone and trumpet mouthpiece as a function of frequency, calculated with (28.2).

The effective length (for an equivalent simple cylinder) of a flared horn increases with frequency. Compared to a simple cylinder, the horn raises the frequencies of the lower modes more, as these tend to stay within the cylindrical tube before the bell, whereas at high modes there is more penetration of the waves into the horn region [8] (see Fig. 28.3), and therefore less percentage increase in frequencies.

An important quantity for wind instruments is the acoustic impedance Z_a, defined as the ratio of sound pressure p to volume velocity U, the latter being the amount of air flowing across the cross-section area of an element per second as the sound wave passes. A plot of Z_a versus frequency f has characteristic peaks at the excitation frequencies of the system and is able to differentiate clearly between, say, a trumpet with and without the mouthpiece [2,3].

Goals and possibilities

- The trumpet is a tube of length 1.4 m and internal diameter 1.1 cm. Attach a rubber membrane to a straight tube of this length and diameter and excite the membrane with a loudspeaker. Show with a microphone that the resonances are those of a closed tube. A skilled trumpet player could excite these resonances by blowing into the tube.

- Insert a trumpet mouthpiece into this tube and note the resonances. Compare them with those of a trumpet. The lowest resonance of the trumpet is most easily excited by replacing the trumpet mouthpiece with a clarinet mouthpiece using a suitable adaptor.

- Have a trumpet player play a trumpet or trombone so as to produce all the notes he can with one slide position or fingering. Then remove the bell and repeat. Record and analyze the sounds with a wave analyzer to show that without the bell the resonances are *not* a harmonic series (out of tune).

- Because of the bell, the radiated spectrum of sound is determined by both the internal spectrum of standing waves and the radiation characteristics of the bell. Record and analyze the radiated sound reaching a microphone say 0.5 m from the horn, and at the same time insert a miniature microphone inside the instrument to probe the inside resonance. This will allow a comparison of the two spectra and lead to further understanding of the bell's role.

- A series of measurements may be made showing how gradual insertion of a player's hand into the bell (a technique used in the French horn) changes the number of resonances and intensity of sound.
- Small objects like a match may be inserted into various parts of the trumpet's air column, in order to show shifts in frequency depending on the position of the object relative to the standing wave nodes or antinodes. Some notes thus produced may be made clearer, others spoilt.
- The spectrum of these instruments is a function of both the pitch and of the loudness. An increase in sound level activates higher resonances, and in addition the radiation efficiency of the bell is heavily frequency dependent. It is therefore of interest to play the instruments at various loudness levels, then record, analyze, and understand the spectrum.
- One could explore the pressure level and distribution in the bell when a sustained note is played: a small microphone could be inserted into the bell in a controlled manner in order to see how far the standing wave penetrates into it.
- Brass instruments are curved so as to restrict their lengths. Too-sharp curves would certainly affect frequencies, as the flared tube does, say if the radius of bending were comparable to the tube radius. Some experimenters might well make a thorough investigation of this bending effect on the sound spectrum of a tube of given length and diameter.
- A major, research-grade undertaking, definitely for the intrepid, and needing the support of a mechanical workshop and skilled technicians, would be to generate experimentally the acoustic impedance curves for pipes and

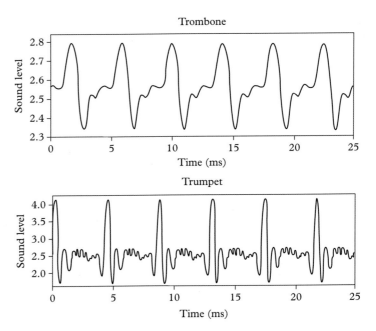

Figure 28.2 *Waveforms for the trombone and the trumpet, reproduced from [8].*

the instruments, with or without the bell [2,3]. This requires a driving mechanism to replace the lips, and a controlled airflow at constant velocity from a driven loudspeaker coupled to the trumpet, so that the pressure response of a small microphone at the mouthpiece will be proportional to the acoustic impedance.

Experimental suggestions

The quantities S and L_e in (28.1) are not hard to measure. The volume of the cup and the tapered tube joined to it may be measured by filling it with water, using a graduated cylinder. Then V is calculated by subtracting an estimate of the tapered tube's volume.

In order to use (28.2) and plot L as a function of f, one takes f as the playing frequency, obtained by a microphone connected to an oscilloscope when playing various notes. The waveform will give f, or it may be obtained from the FFT of the signal. It will be interesting to see whether the frequencies form a harmonic series. Comparison of internal and external sound spectra is best done with a two-channel spectrum analyzer.

Use thin-walled brass tubing of internal diameter 7/16″ (11 mm) for experiments on tubes with and without bells and/or mouthpiece. This will be a fair approximation to the tubes used in trumpets. Various obstacles, say a finger or a rod, may be inserted into the bell, in order to see how a small microphone connected to an oscilloscope records the shift of the turning point where the wave is reflected in the bell. Finally, here are some data, reproduced from [8], for various characteristics of the trombone and the trumpet. Figure 28.2 shows waveforms for the tones produced.

Figure 28.3, reproduced from [5,6,8], shows how the higher modes penetrate into the bell. The lower modes tend to stay in the cylindrical sections of the instruments, resulting in a shortened air column and therefore higher frequencies.

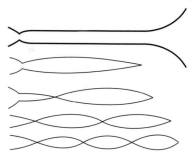

Figure 28.3 *Mode penetration into bell.*

REFERENCES

1. H. von Helmholtz, *On the Sensations of Tone* (Dover, New York, 1954).
2. A.H. Benade, *Fundamentals of Musical Acoustics* (Dover, New York, 1990).
3. A.H. Benade, The physics of brasses, *Sci. Amer.* **229**(1), 24 (1973).
4. J. Backus, *The Acoustical Foundations of Music* (Norton, New York, 1977), ch. 12.
5. T.D. Rossing, F.R. Moore, and P.A. Wheeler, *The Science of Sound* (Addison-Wesley, New York, 2002).
6. N.H. Fletcher and T.D. Rossing, *The Physics of Musical Instruments* (Springer, New York, 1998).
7. R.W. Pyle, Effective length of horns, *J. Acoust. Soc. Am.* **57**, 1313 (1975).
8. M.C. LoPresto, Experimenting with brass musical instruments, *Phys. Educ.* **38**, 300 (2003).

Part 4

Liquids

29	Sound from gas bubbles in a liquid	163
30	Shape and path of air bubbles in a liquid	168
31	Ink diffusion in water	173
32	Refractive index gradients	176
33	Light scattering by surface ripples	180
34	Diffraction of light by ultrasonic waves in liquids	184
35	The circular hydraulic jump	188
36	Vortex physics	192
37	Plastic bottle oscillator	197
38	Salt water oscillator	201

Part 6

Liquids

Sound from gas bubbles in a liquid

29

Introduction

The murmur of the brook, the roar of the cataract, or the noise of the surf find partial explanation in the sound emitted into the surrounding water by expanding and contracting gas bubbles of various diameters [1]. This project aims to study in detail the amplitude and damping of oscillating air bubbles in water, but may be extended to bubbles filled with other gases and surrounded by different liquids. Both laboratory models and actual sounds of brooks in situ may be recorded and analyzed, granting the investigation the charm of being directly connected with a natural phenomenon. We summarize here work done 80 [1] and 30 [2] years ago.

Theoretical ideas

Consider a gas bubble "breathing" with amplitude A around a mean radius R while it is surrounded by an incompressible liquid of density ρ extending to distances many orders of magnitude larger than R, as shown in the Fig. 29.1.

If the oscillations around R are simple harmonic, $A\sin(\omega t)$, the natural frequency of the bubble can be calculated from energy considerations, together with the equation of state of the enclosed gas. Since a priori the process need be neither isothermal nor adiabatic, one may only assume that the gas pressure p in a bubble and its volume V are related by $pV^\kappa = $ constant, where κ is a parameter to be determined experimentally, its value lying between unity and γ, the ratio of specific heats.

Let r be any radius and $\dot{r}$ its time derivative. When the bubble has radius r, the liquid has kinetic energy E_k given by $\int_r^\infty \frac{1}{2}(4\pi r'^2 \rho)\dot{r}'^2 dr'$. Here, ρ is the density, and r' the integration variable. Assuming that the liquid is incompressible, we have $\dot{r}'/\dot{r} = r^2/r'^2$, and therefore the kinetic energy integral is equal to $2\pi \rho r^3 \dot{r}^2$. The maximum value of this energy, when $r = R$ and $\dot{r} = A\omega$, is $E_k = \frac{1}{2}(4\pi R^3 \rho)(A\omega)^2$. The quantity $4\pi R^3 \rho$ is the effective (radiation) mass of the liquid. This energy is then equated to the potential energy at the extreme radius of motion, which is equal to the work done in compressing the bubble from radius R to radius $(R - A)$, namely, $6\pi\kappa p_1 RA^2$. Here, p_1 is the total pressure in the surrounding

Introduction	163
Theoretical ideas	163
Goals and possibilities	164
Experimental suggestions	165

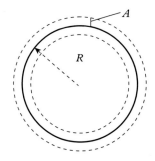

Figure 29.1 *Oscillating gas bubble.*

liquid, comprising the sum of the atmospheric and pressure-head contributions. Hence the natural frequency of oscillation of the bubble is given by

$$f = (3\kappa p_1/\rho)^{1/2}/2\pi R. \tag{29.1}$$

For millimeter-radius air bubbles in water, f should be in the kHz range.

Any instrument recording the sound from such bubbles will register the sound pressure $P(r,t)$, whose amplitude P_0 is related to the oscillation amplitude A and the wave number k by [3]:

$$P_0 = \frac{A(2\pi f)^2 \rho R}{(1 + (kr)^2)^{1/2}}. \tag{29.2}$$

This expression enables one to estimate the oscillation amplitude in terms of the recorded pressure amplitude.

It is expected that the oscillations of the bubble will be damped. The damped harmonic motion can be represented by a damped harmonic differential equation, and from it one can extract the decay constant Γ and the quality factor Q, in the usual way. There are three possible loss mechanisms:

(a) Viscous damping. For bubbles of 1 mm or less this has been shown to be negligible [4].

(b) Acoustic radiation resistance [3], given by

$$R_{\text{ac.rad}} = 4\pi R^2 \rho c (kR)^2, \tag{29.3}$$

where c is the velocity of sound.

(c) Thermal radiation resistance, due to the heat produced on compressing the bubble [4]:

$$R_{\text{therm.rad}} = D(4\pi R^3 \rho)(2\pi f)^{3/2}, \tag{29.4}$$

where the constant $D = 1.6 \cdot 10^{-4} s^{1/2}$ for bubbles of the order of 1 mm. Given these losses, one may calculate the quality factor Q and the energy loss per cycle $2\pi/Q$:

$$Q = 1/[(\omega R/c) + D\omega^{1/2}]. \tag{29.5}$$

Goals and possibilities

- Measure the period of oscillation of a single bubble released into a liquid, say air bubbles into water.
- Vary the bubble size, the enclosed gas, and the surrounding liquid, in order to test the Minnaert frequency formula in (29.1).

- Measure the maximum oscillation amplitude.
- Measure the decay time, the quality factor, and the energy loss per cycle for oscillating bubbles in different cases and compare with theoretical estimates.
- Record the sound emitted by a brook in an area which is clearly bubble generating, and analyze the frequency spectrum. From this obtain an idea of the distribution of bubble size in that part of the stream.

Experimental suggestions

See Ref. [2]. A possible laboratory setup is shown in Fig. 29.2.

The lucite container can be 20 × 20 × 40 cm. Sound produced by an oscillating bubble released from the nozzle is picked up by a hydrophone, amplified, stored in a digital oscilloscope, and passed to a wave analyzer. A piezoelectric waterproofed loudspeaker may be substituted for the hydrophone. It is necessary to calibrate the microphone used, using any of the numerous available methods [5,6].

Bubble diameter is controlled by using various Pasteur pipettes of suitable diameters, drawn out from thin glass tubing. Bubbles are released singly from a syringe controlled by a micrometer attached to and pushing its handle. Average bubble volume may be estimated by measuring the total volume of gas collected in an inverted liquid-filled cylinder when a large number of bubbles are released into it. A three-way tap (not shown in Fig. 29.2) allows the release of bubbles and filling of the syringe with different gases.

Record the sound of bubbling brooks by immersing the hydrophone in a suitable place, recording the sound on a tape recorder, and then playing it back into

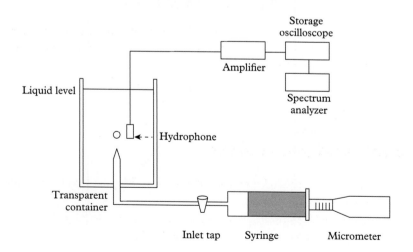

Figure 29.2 *Apparatus for recording sound from bubbles.*

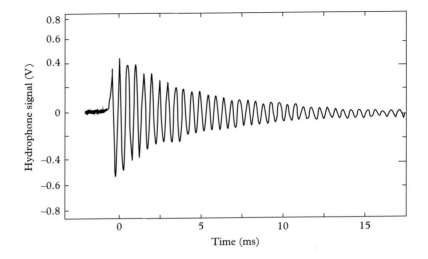

Figure 29.3 *Hydrophone signal from a bubble, reproduced from [2].*

a storage oscilloscope, and subsequently a wave analyzer. The spectrum of frequencies in the recording will give an idea, via the frequency formula (29.1), of the bubble-size distribution in the brook.

Figure 29.3, reproduced from [2], shows a sample of the hydrophone output due to the sound emitted from an air-filled bubble of radius 1.7 mm in water.

Figure 29.4 shows the time dependence of the pressure $P_0(t)$ for the trace in Fig. 29.2.

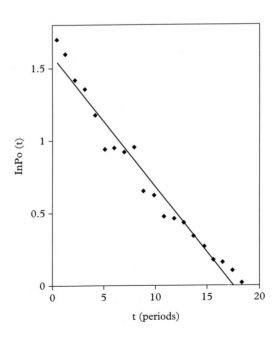

Figure 29.4 *A plot of $ln(P_0)$ versus time for the signal in Fig. 29.3.*

REFERENCES

1. M. Minnaert, On musical air-bubbles and the sounds of running water, *Phil. Mag.* **16**, 235 (1933).
2. T.G. Leighton and A.J. Walton, An experimental study of the sound emitted from gas bubbles in a liquid, *Eur. J. Phys.* **8**, 98 (1987).
3. L.E. Kinsler and A.R. Frey, *Fundamentals of Acoustics* (Wiley, New York, 1980), and later editions.
4. C. Devin, Survey of thermal radiation and viscous damping of pulsating bubbles in water, *J. Acoust. Soc. Am.* **31**, 1654 (1959).
5. J. Wren, How to calibrate a microphone, 2011, video available online at <http://www.youtube.com/watch?v=GsF1jnTjRLo>.
6. V. Nedzelnitsky: Laboratory microphone calibration methods at the National Institute of Standards and Technology, available online at <http://www.nist.gov/calibrations/upload/aip-ch8.pdf>.

30 Shape and path of air bubbles in a liquid

Introduction 168
Theoretical ideas 168
Goals and possibilities 169
Experimental suggestions 170

Introduction

The motion of air bubbles rising in various fluids has been studied intensively over the last century. The problem attracts much attention, because bubble–fluid interactions are important in such diverse fields as air-sea transfer, bubble column reactors, oil/gas transport, boiling heat transfer, ship hydrodynamics, inkjet printing, and medical ultrasound imaging.

The subject is an area of active research, focusing on a number of problems. To name but a few: the shape of a bubble as a function of the dimensionless Weber number; the role of surfactants on the drag force exerted on a bubble; how drag varies with the aspect ratio (the ratio of the major to minor axis) of an ellipsoidal bubble; whether and under what conditions the path of a bubble is rectilinear, zigzag, or spiral; and how its terminal velocity may be dependent on the bubble dimensions. Some of these questions may be addressed in an advanced undergraduate project as outlined below.

Theoretical ideas

A full understanding of the theory requires advanced knowledge of boundary layer theory, and irrotational and rotational viscous hydrodynamics. Therefore this project should be carried out only under the guidance of an expert in hydrodynamics, who will be able to explain and justify the various assumptions outlined below. We refer the reader to Refs [1–5] for details and will only state some assumptions which have relevance to the kind of experimentation carried out in this project.

In the theory [3], shown to be valid for many volatile organic liquids and water, the viscosity of the liquid is taken to be small, and surface tension plays an important part through the pressure field generated across the bubble. It is also assumed that the bubble surface cannot support any tangential stress. This assumption does not hold hold in the presence of surfactants [6]. In the latter case molecules of the impurity collect at the bubble surface as it ascends, and the bubble will behave like a small solid body. The drag coefficient C_D is calculated as an expansion in inverse powers of the Reynolds number $R = a\rho U/\mu$, where a, ρ, U, and μ are the bubble radius, liquid density, rise speed of the bubble, and liquid viscosity, respectively; at low Reynolds numbers it is inversely proportional [7] to R. The drag coefficient is found to rise steeply as the ratio of the major to minor axis (aspect ratio) of the bubble increases.

The shape of the bubble is determined by the pressure field of the irrotational flow. It has been shown that a symmetric ellipsoidal shape is possible provided that the Weber number, defined as the ratio of the dynamic pressure ρU^2 to the surface tension pressure σ/a (where σ is the surface tension and a is the radius of an equivalent sphere having the same volume as the bubble), is of order unity. For a class of "thin" liquids like water and many volatile organic liquids, distortion from a spherical shape sets in at a Reynolds number of about 100.

A quantity of interest, both theoretically and experimentally, is the variation of the terminal velocity U of a bubble as a function of a. Figure 30.1, reproduced from [3], shows some typical experimental results and comparison with theory [5,8]. (Note: in [3], R rather than a denotes the radius.)

A further interesting phenomenon is the transition from rectilinear to zigzag paths for spherical bubbles, and from rectilinear to spiral paths for ellipsoidal bubbles, as the effective radius is increased [9]. Some typical results are shown in Fig. 30.2, reproduced from [8].

Goals and possibilities

- Build a large enough vertical tank so that the motion of bubbles in the liquid will not be influenced by the walls.
- Devise a way to release bubbles of different sizes and shapes, both spherical and ellipsoidal, from capillary tubes of different diameters, fixed in position

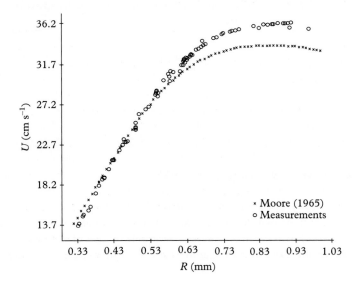

Figure 30.1 *Vertical rise velocity of a bubble in pure water as a function of radius R.*

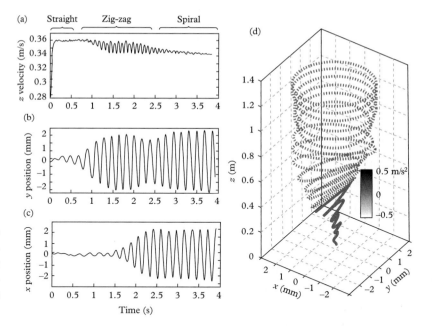

Figure 30.2 *Trajectory of a 1.2 mm bubble: (a) vertical component of velocity; (b) y position from camera data; (c) x position; (d) three-dimensional reconstruction of full trajectory – grayscale indicates magnitude of acceleration.*

at the bottom of the tank. Concentrate on effective bubble radii larger than 0.9 mm, for which surfactant effects are negligible, so that one may use tap water rather than less available highly purified water appropriate for research-grade investigations.

- Follow the vertical rise of the bubble by any of the means reported in the literature: this could involve use of a high speed video camera [8], or using ultrasonic waves bounced off the bubble to determine its speed from the Doppler-shifted frequency of the ultrasound [9].

- Try to follow the transition from rectilinear to zigzag and to helical paths in the course of the vertical rise, as the radius of the bubbles is increased from 0.9 mm towards 1.5 mm. The height at which the transition will occur depends on the size of the bubble.

- Measure the terminal velocity of the bubbles as function of the radius and compare the results with Moore's theory [3].

Experimental suggestions

Obtain a series of capillary tubes in the 0.9–1.5 mm range.

A number of methods have been used for producing bubbles. In one of them [9], air is pumped through a stainless steel capillary tube of 0.3 mm diameter by connecting it with Tygon tubing to a peristaltic pump turned by hand. The tube faces with its open end vertically upwards. Alternatively [8], one may use small syringes controlled by syringe pumps, pumping air into a short teflon tube of similar radius, at a very slow rate, in order to ensure quasistatic release of the bubbles into the water.

It has been shown [8] that when the capillary internal diameter is about the same as that of the bubble (0.12 cm), the latter is released in spherical form, whereas when the capillary diameter is much smaller (0.03 cm), the bubbles tend to be ellipsoidal. This should be kept in mind when choosing the capillary.

The volume of a bubble may be determined by trapping it at the end of its rise, sucking it into a thin transparent tube of known cross section with a syringe, with water on either side of the bubble. The length of the air plug will then give the volume and from it the effective radius r_{eff} may be calculated. Figure 30.3 sketches the apparatus. An ultrasonic emitter positioned at the top of the container generates a continuous downward beam, operated at 2.8 MHz. The sound is scattered by the rising bubble, its frequency is Doppler shifted, and it returns to the top where it is detected by another ultrasound module. The frequency shift is proportional to the bubble rise velocity U. It is a non-trivial matter to amplify, digitize, and Fourier transform the signal. Such routines are available in a number of commercial software packages, Matlab® among them.

The lateral motion of the bubble may be recorded by a high speed video camera, situated on top of the reservoir, near the ultrasound modules. The available budget will determine the accuracy of recording: 60 frames/s is quite common with digital cameras, 125 frames/s or more approaches research-grade equipment. Image processing routines, many freely available on the internet (ImageJ 1.4 being an example), or in Matlab, will be able to extract the lateral position as a function of time. From this, lateral velocities may be obtained by numerical differentiation.

Clearly, the heart of the project consists of obtaining the trajectory and velocities from the signal and image processing of the recording equipment.

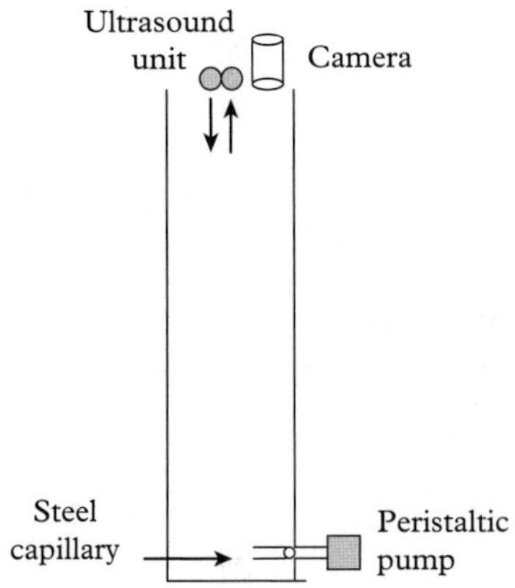

Figure 30.3 *Producing and tracking bubbles.*

REFERENCES

1. V.G. Levich, *Physico-Chemical Hydrodynamics* (Prentice Hall, 1962).
2. V.G. Levich, Motion of gaseous bubbles with high Reynolds numbers, *Zh. Eksp. Teor. Fiz.* **19**, 18 (1949).
3. D.W. Moore, The velocity of rise of distorted gas bubbles in a liquid of small viscosity, *J. Fluid Mech.* **23**, 749 (1965).
4. J. Magnaudet and I. Eames, The motion of high-Reynolds-number bubbles in inhomogenous flows, *Annual Rev. Fluid Mech.* **32**, 659 (2000).
5. P.C. Duineveld, The rise velocity and shape of bubbles in pure water at high Reynolds number, *J. Fluid Mech.* **292**, 325 (1995).

6. M. Hameed et al., Influence of insoluble surfactant on the deformation and breakup of a bubble in a viscous fluid, *J. Fluid. Mech.* **594**, 307 (2008).
7. Wikipedia, Drag coefficient, online at <http://en.wikipedia.org/wiki/Drag_coefficient>.
8. M. Wu and M. Gharib, Experimental studies on the shape and path of small air bubbles rising in clean water, *Phys. Fluids* **14**, L49 (2002).
9. W.L. Shew, S. Poncet, and J-F. Pinton, Force measurements on a rising bubble, *J. Fluid Mech.* **569**, 51 (2006).

Ink diffusion in water

31

Introduction

Although diffusion phenomena are of great importance in technology, physical chemistry, biology, and solid state physics, they are rarely included in the high school or undergraduate physics curriculum, except perhaps as examples of the solution of partial differential equations. Through the present project students will be exposed to the topic and carry out an experimental study to reinforce their understanding of diffusion.

Introduction	173
Theoretical ideas	173
Goals and possibilities	174
Experimental suggestions	174
Further explorations	175

Theoretical ideas

We consider the spreading of a drop of ink, the solute, dropped into the center of a shallow circular dish of water, the solvent [1]. The physics is then well described by a diffusion equation based on Fick's law, namely, that the rate of transfer is proportional to the concentration gradient. In the two-dimensional system shown in Fig. 31.1 possessing circular symmetry, the polar coordinate form of this equation is

$$D\left(\frac{\partial^2 c}{\partial r^2} + \frac{1}{r}\frac{\partial c}{\partial r}\right) = \frac{\partial c}{\partial t}, \tag{31.1}$$

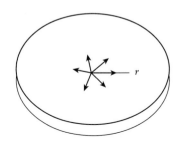

Figure 31.1 *Drop diffusion.*

where D is the diffusion constant of ink in water and $c(r,t)$ is the concentration of ink. This equation is solved subject to the initial conditions. We assume that at $t = 0$ the concentration is a constant c_0 in a region of size $r = a$, where a the radius of the ink drop, and $c = 0$ for $r > a$. Then for a vessel of infinite extent the solution of (31.1) is obtained as follows:

$$c(r,t) = \frac{c_0}{2Dt} \exp\left(-\frac{r^2}{4Dt}\right) \int_0^a \exp\left(-\frac{r'^2}{4Dt}\right) I_0\left(\frac{rr'}{2Dt}\right) r' dr', \tag{31.2}$$

where I_0 is the modified Bessel function of order zero. This function $c(r,t)$ may be evaluated by numerical integration for c/c_0 as a function of r/a in units of the diffusion time a^2/D, and serves as the basis for the experimental measurements.

Goals and possibilities

- Evaluate (31.2) numerically, plot c/c_0 as function of r/a for various times expressed in units of a^2/D and confirm the following qualitative conclusions:

    (a) at $r = 0$ there is no concentration gradient for all t;

    (b) c is a maximum at $r = 0$ for all t;

    (c) c decreases with r for any particular t;

    (d) for $r \leq a$ the concentration c is a monotonically decreasing function of t.

- Measure the spread of the ink front as function of time and compare with the theoretical predictions.
- Repeat the experiment for different temperatures.
- Extract the diffusion coefficient D from the measurements and its dependence on temperature T.
- By plotting $\ln(D)$ as function of $1/T$, check whether the Arrhenius relation $D = A\exp(-E/k_B T)$ relating the diffusion constant D to the activation energy E is satisfied. Here, T is the absolute temperature and k_B Boltzmann's constant. From the plot extract the activation energy E, the energy barrier for an atom of ink to diffuse.
- Try the experiment with a different ink, measure the viscosities of the two inks, and relate these to the magnitudes of the respective diffusion coefficients.

Experimental suggestions

Use a large flat transparent circular dish filled with water to a depth of only 1 cm, in order to cancel any gravity-influenced spreading of the ink drop. The larger the dish, the less the influence of its boundary on the diffusion process, and the better the expected agreement with (31.2), which was obtained for an infinite dish, for large times. Keep the dish at a constant temperature by immersing it in a thermostat-controlled water bath. This is often available in biology laboratories.

Fountain-pen ink may be injected into the center of the dish using a hand-drawn glass tube or pipette. Its spread with time may be continuously recorded by a video camera fixed directly above the dish and then analyzed frame by frame, or by a regular digital camera every few seconds. A suitable radial distance scale needs to be established by previous calibration. This can be done by fitting a fine polar diagram over the dish and taking a photo of it.

From the measurements, the radius of the ink front as function of time may be plotted and compared with the theoretical prediction from (31.2).

Observations may be repeated at different temperatures, covering the available range of the water bath temperature. This enables one to extract the diffusion constant as a function of temperature T, and to obtain an Arrhenius plot of $\ln(D)$ versus $1/T$, expected to be a straight line, from which one can determine experimentally the activation energy E for the ink.

Since diffusion depends on the viscosity η, and since the latter is a sensitive function of temperature, good control of the latter is vital.

Measuring the viscosity η of the various inks used would extend the demands on apparatus considerably, unless a commercial viscometer is at the experimenter's disposal. In that case she could correlate experimentally η with D and T, in order to see whether the Stokes–Einstein relation $D = kT/6\pi\eta r$ is satisfied. Here, r is the radius of the diffusing particle.

Further explorations

References [2,3] suggest additional, more challenging, experimental projects on diffusion, which will deepen the understanding of the subject. They will require more sophisticated instrumentation and so are only suitable in a research-grade environment.

....

REFERENCES

1. S. Lee, H. Lee, I. Lee, and C. Tseng, Ink diffusion in water, *Eur. J. Phys.* **25**, 331 (2004).
2. D.R. Spiegel and S. Tuli, Transient diffraction grating measurements of molecular diffusion, *Am. J. Phys.* **79**, 747 (2011).
3. B. Clifford and E.I. Ochiai, A practical and convenient diffusion apparatus: An undergraduate physical chemistry experiment, *J. Chem. Educ.* **57**, 678 (1980). Also available online at <http://www.pubs.acs.org/toc/jceda8/57/9>.

32

Refractive index gradients

Introduction 176
Theoretical ideas 176
Goals and possibilities 177
Experimental suggestions 178

Introduction

Mirages, shimmering air above a hot road surface or around a candle flame, and sound getting "lost" in the atmosphere are just a few examples of wave propagation in media with optical or mass density gradients. Often the bending of light in an inhomogeneous medium is demonstrated by passing it through a container in which diffusion establishes a solution with a vertical refractive index gradient. The present project measures the refractive index n and its first and second derivatives as functions of vertical distance y, and the variation of these quantities with time. This can lead to a determination of the diffusion coefficient for a given pair of liquids. Several methods can be used to measure $n(y)$, requiring the researcher to learn some interesting physics [1,2].

Theoretical ideas

Imagine light refracted through a cell containing a sugar solution whose refractive index varies continuously with the vertical coordinate y – see Fig. 32.1. The width of the cell is a. This medium may be represented by a multi-layered liquid, each of which has an increment dn in its refractive index relative to the one above it. By Snell's law, the quantity $n(y)\sin\theta_y$ is a constant at any layer interface, where θ_y is the angle to the normal. This can be shown by considering refraction of neighboring thin layers as the thickness becomes very small, or by using Fermat's principle.

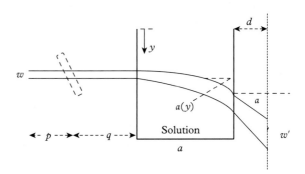

Figure 32.1 *Refraction through a cell.*

Suppose now that, after passing through the medium, an initially horizontal beam of light exits at angle α to the incident direction. Then for small angles of deviation we may show that

$$\alpha = a(dn/dy), \qquad (32.1)$$

so that measuring α will give dn/dy.

Let the incident beam have a width w. Then because of the spatial variation of n, the exiting beam has a width w', and the two are related by the geometry of the system and the second derivative of the refractive index d^2n/dy^2, which can therefore be measured. For small angles we find (refer again to Fig. 32.1 for the symbols),

$$w' = w(da)d^2n/dy^2. \qquad (32.2)$$

The function $n(y)$ itself may be measured in a number of ways, as described in the following sections, where we provide further theoretical relationships relevant for each method.

Goals and possibilities

- Determine $n(y)$ at any y by measuring the real and apparent thickness of the cell by viewing its back wall through a traveling microscope at the appropriate height, using the formula $n = $ (real depth)/(apparent depth).
- Measure $n(y)$ by measuring the *osmotic pressure* in the solution at any y. The index of refraction is proportional to the mass concentration of sugar $c(y)$ (sugar density as a function of height, in g/cm^3), which itself depends on the osmotic pressure. This method will require calibration with a solution of *known* concentration.
- The concentration itself may be determined by measuring the decrease in intensity between the incident and transmitted beam. This decrease is proportional to the concentration. Once again, calibration is necessary with a solution of known concentration.
- Fit the measured values of $n(y)$ for different y with a single function to represent it throughout the solution.
- Measure dn/dy as outlined in the theory section, integrate it to obtain $n(y)$, and compare it with the proposed functional form for $n(y)$.
- Study the mixing zone refractive index as a function of time, in order to extract information on the diffusion constant, as shown by Sommerfeld [3].
- Different pairs of liquids may be studied, each intermingling at different rates and therefore having different diffusion constants.

Experimental suggestions

Look for, or make, a $8 \times 20 \times 2$ cm transparent container (the cell) with as thin walls as possible. Use a He–Ne laser for a well-defined beam, projected perpendicular to the 8×20 face. Set the cell on a laboratory jack, so as to be able to pass a light beam through it at any height without disturbing the solution in it. Alternatively, pass the beam through a cylindrical glass rod 5–6 mm in diameter, placed at 45° to the beam path, and adjust the distances p and q (see Fig. 32.1) so that the expanded beam covers much of the cell's depth. Half-fill the cell with water, then slowly pour a sugar solution (80 g sugar, 50 ml water) through a funnel down a thin plastic pipe fastened to one wall and reaching to the bottom, regulating the flow with a clip on the pipe. Observe the apparent beam width on a screen located at a convenient distance from the jack. Sugar solution and water have a diffusion rate which need to be observed over a time period of several days.

Other liquid pairs may be used. Glycerin and water [2] will mix much faster, allowing diffusion effects to be observed in a couple of hours. The apparent width of the container can be measured accurately at any height by using a traveling microscope, viewing and focusing on the back wall normal to the width. The effective beam width of the laser before and after refraction can be measured with a ruler or with more accurate methods (intensity measurements by a photodetector).

Osmotic pressure measurements require a thin glass tube, one end of which is closed with a thin semipermeable membrane. The latter is held vertically in the solution at the height y below the upper level at which measurement is desired. Water will penetrate the membrane and be held to height H in the tube by the osmotic pressure (see Fig. 32.2). Let ρ_a, ρ_w, k_B, T, and M denote the average density of the solution, the density of water, Boltzmann's constant, the temperature, and the mass of the sugar molecule, respectively. Then for dilute solutions:

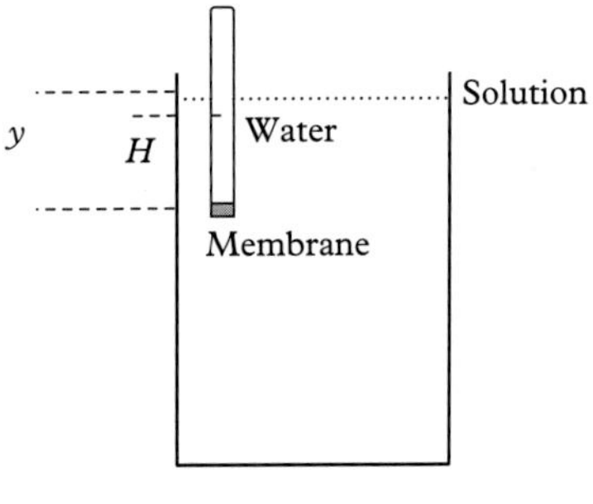

$$c(y) = M(\rho_a g y - \rho_w g H)/k_B T. \qquad (32.3)$$

Figure 32.2 *Osmotic pressure.*

The quantity in brackets is the difference in pressure in the solution at height y and that in the tube just above the membrane.

The angle of rotation $\theta(y)$ of the beam's plane of polarization on passing through the solution at any height y can be measured by using crossed polarizers mounted on rotating holders graduated in degrees, whereby one polarizer is placed into the beam's path before, the other after, the container: Obtain the minimum intensity with and without the solution interposed. For dilute solutions $c(y) = b\theta(y)$, so that knowing b would furnish another method for getting $n(y)$ experimentally. For a solution of *known* uniform concentration, measuring n and θ will determine b.

For measuring the diffusion constant it is better to use a smaller cell, say $5 \times 5 \times 0.6$ cm, so that all of its vertical face may be covered by the expanded light beam. One then looks for the ray deflected by the greatest angle α_{max} on passing through the cell, as a function of time t. Sommerfeld [3] has shown that

$$\alpha_{max} = (n_2 - n_1)a/2\sqrt{\pi Dt}. \tag{32.4}$$

Here, D is the diffusion coefficient and n_2, n_1 are the refractive indices for the bottom and top liquids, respectively.

....

REFERENCES

1. A.J. Barnard and B. Ahlborn, Measurement of refractive index gradients by deflection of a laser beam, *Am. J. Phys.* **43**, 572 (1975).
2. D.A. Krueger, Spatially varying index of refraction, *Am. J. Phys.* **48**, 183 (1980).
3. A. Sommerfeld, *Optics* (Academic Press, New York, 1952), pp. 347–50.

33 Light scattering by surface ripples

Introduction 180
Theoretical ideas 180
Goals and possibilities 181
Experimental suggestions 182

Introduction

Single- and multiple-slit interference and diffraction experiments are usually performed by students studying wave phenomena. But carrying out scattering or diffraction experiments with other arrays like crystals or density waves requires expensive apparatus which is seldom available.

In the following we propose a project that can be performed with limited resources, involves some planning and construction by the researchers at varying levels of sophistication, and can take anything from several days to a term. The idea is to scatter laser light from surface waves on a liquid, generated by a vibrator which is controlled by a continuously variable audio frequency generator with a power output. Several aspects of the project require initiatives: the building of probes to generate circular and plane waves, looking at spatial coherence via fringe visibility, designing a mechanically stable apparatus, and the choice of photodetection devices. The theory of surface tension waves serves as a necessary background.

Theoretical ideas

Since the usual study of reflection gratings is for a plane array, it is natural to generate plane wave ripples with an appropriate probe. The scattered light amplitude would at first sight be just the Fourier transform of the surface wave displacement. But the system is more subtle and richer. The aperture function is controlled by the phase changes of the scattered wave caused by variations of the surface height h. In addition, wave crests may completely block reflections. So the diffraction envelope is affected both by the ripple amplitude h, as well as the frequency ω. Thus, some maxima may be diminished in intensity, others may be missing.

The angular position of the diffraction is given as usual in terms of the optical path difference of the diffracted light from neighboring ripples. Let the laser be incident at an angle θ to the liquid surface, k the wavenumber of the ripples, λ the laser wavelength, θ_n the nth-order spot direction, as in Fig. 33.1. Then

$$(2\pi/k)(\cos\theta - \cos\theta_n) = n\lambda. \tag{33.1}$$

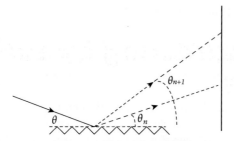

Figure 33.1 *Geometry for diffraction spots.*

This enables one to calculate the angular spacing of the spots $\Delta_n = \theta_{n+1} - \theta_n$, which depends on n. If $\Delta_n/\theta \ll 1$ we get

$$\Delta_n = \lambda k / 2\pi \sin \theta_n. \tag{33.2}$$

If our probe were point-like and produced circular waves, we would expect the same formulae to apply, provided we observe the pattern at a distance that is many ripple wavelengths away, and θ is small, so that the wavefront curvature plays a negligible role.

The dispersion relation for surface tension waves (gravity and viscosity are neglected) is given by [1]

$$\omega = (\sigma/\rho)^{1/2} k^{3/2}, \tag{33.3}$$

where σ is the surface tension and ρ is the density of the liquid. Measurement of ω and k experimentally can test this relation. This is also a neat way to measure the surface tension of various liquids.

Goals and possibilities

- Find the angular spacing of the diffraction spots as a function of angle.
- Explore the influence of a gradually increasing ripple amplitude on the intensity and number of diffraction spots.
- Look at the effect of reflecting the beam near to and far from the probe area where the wave period influences the stability of the diffraction pattern (since the ripples spread apart there).
- Establish the validity of the dispersion relation and obtain the ratio σ/ρ for various liquids.
- Quantify the spot intensities and fringe visibility with a photodetection system.
- Experiment with various liquids, such as water, ethanol, or their mixtures, and determine experimentally their surface tension, both as a function of concentration and as a function of temperature.

Experimental suggestions

Stability of the ripple pattern is essential. The vessel for the liquid should be supported on a flat stone or cement slab resting on inflated tire inner tubes. Draughts are to be excluded from the area. Clean liquids and grease-free containers are vital.

The vibrator could be from a commercial ripple tank apparatus, but these are expensive. One could glue the probe to the diaphragm of a loudspeaker driven by a sine wave audio signal generator with a variable 0–6 V low-impedance output.

The probe can be needle-like or gate-like, the latter say in the shape of a "square well" from thin stiff wire, as shown in Fig. 33.2. The probe depth into the liquid surface must be adjusted to avoid splashing and non-linear effects, either by raising the probe or adjusting the vessel's height with a lab jack. The liquid depth should be around 1–2 cm in order for the above dispersion relation to hold.

The signal frequency can be measured by a frequency meter. Modern multimeters are equipped to do this. For each driving frequency the loudspeaker amplitude is an interesting parameter, but care must be exercised in order to avoid the generation of audible harmonics when the amplitude is too high.

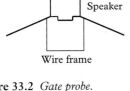

Figure 33.2 *Gate probe.*

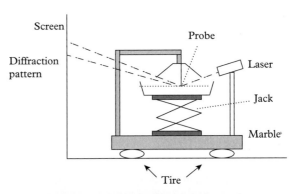

Figure 33.3 *Experimental setup.*

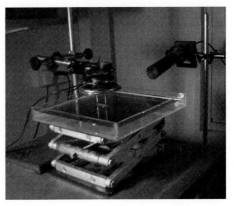

Figure 33.4 *Apparatus for light scattering on ripples.*

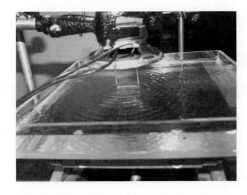

Figure 33.5 *Generation of ripples.*

The position of the diffraction spots must be measured accurately, since they help determine k, which enters the dispersion relation raised to a power. A photodiode or other sensor movable on a vertical axis could measure spot intensities and quantify some of the investigations suggested above. Its position could be determined by a motion sensor connected to a computer. A schematic of the setup and a photograph of the apparatus are shown in Figs 33.3 and 33.4.

Figure 33.5 shows the ripples generated, and Fig. 33.6 is a typical interference pattern which gives information on the wavelength of the ripples generated by the wire.

To measure the temperature dependence of the surface tension you could heat the liquid to 100°C and then let it cool, photographing the diffraction pattern from the ripples, say, every 5°C, while the liquid surface temperature is being monitored by a thermocouple.

REFERENCES

1. I.G. Main, *Vibrations and Waves in Physics*, 3rd ed. (Cambridge University Press, Cambridge, 1994).

Figure 33.6 *Interference pattern.*

34 Diffraction of light by ultrasonic waves in liquids

Introduction 184
Theoretical ideas 184
Goals and possibilities 185
Experimental suggestions 186

Introduction

Two related experimental methods developed in the 1930s are the basis for this project [1,2]. They are based on the diffraction of light when an ultrasonic wave passing through a liquid creates density fluctuations, which serve as a diffraction grating. Both the direct observation of the diffraction pattern, as well as the use of Schlieren methods to view the grating itself, enable one to measure accurately the velocity of sound in the liquid.

Theoretical ideas

Full theoretical calculations are available in the literature [3], both for the angular positions of the light diffraction maxima from a "liquid grating" and for the time-averaged intensity of light in any order, when an ultrasonic wave is excited in a liquid. Perhaps surprisingly, the angle θ_n for the nth-order Fraunhofer diffraction light intensity maximum is given by the usual expression for a ruled grating,

$$d \sin \theta_n = n\lambda, \qquad (34.1)$$

where d, λ are the sound and light wavelengths, respectively. The sound velocity c is found from the value of d, determined experimentally using (34.1), and from the relation $c = \upsilon d$, where υ is the ultrasound frequency.

Unlike for an ordinary ruled grating, the central maximum is not necessarily the brightest, neither is the light intensity given by the usual sinc() function. Instead, the time-averaged light intensity for the nth order, I_n, is given by nth-order Bessel functions $\mathcal{J}_n$. For traveling waves,

$$I_m = \mathcal{J}_m^2(2\pi \ell \delta\mu/\lambda). \qquad (34.2)$$

Here, $\delta\mu = \mu_{\max} - \mu_0$ is the maximum variation of the optical refractive index from its value μ_0 in the absence of sound, and ℓ is the path length of light in the sound medium. For standing waves one finds

$$I_m = \sum_{r=-\infty}^{\infty} \mathcal{J}_r^2\left(\frac{\pi\ell\delta\mu}{\lambda}\right)\mathcal{J}_{r-m}^2\left(\frac{\pi\ell\delta\mu}{\lambda}\right). \qquad (34.3)$$

Both expressions strongly depend on the acoustic intensity via $\delta\mu$. The reader is strongly recommended to study the detailed and somewhat lengthy derivation of these expressions in Ref. [3].

For a qualitative understanding of the diffraction [4] we may assume that the light path in the sound medium is short enough, and the sound intensity low enough, so that the light rays are not curved significantly while passing through the liquid. The light wavefront spreads through the liquid so fast compared to the sound velocity that the sound wavefronts may be taken to form a time-independent (rigid) sinusoidal phase grating in the liquid, with periodicity d. The angular positions of the diffraction maxima will not vary with a transverse displacement of a rigid grating, and for this reason the diffraction pattern produced by a traveling sound wave is like that of a grating at rest.

Consider the basic experimental setup in the Debye–Sears method, sketched in Fig. 34.1. For a standing pressure wave the distance between successive compressions or rarefactions is d, but the places of all such regions oscillate back and forth by a distance $d/2$ at the frequency v of the sound wave. The thick lines in the cell describe compressions at some time t, the thin lines at times $(t+1)/2v$. At intermediate times (say at $(t+1)/4v$, etc.) the liquid density is uniform, causing the diffraction pattern to disappear instantaneously. These changes are undetectable visually, so that one obtains the same pattern for a standing as for a traveling wave.

Instead of observing the Fraunhofer diffraction pattern, one may use shadow techniques in order to observe the striation pattern on a distant screen due to the ultrasonic standing wave grating itself. Details and interpretation are given in the experimental section.

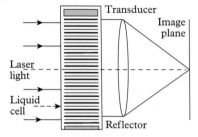

Figure 34.1 *Debye–Sears method.*

Goals and possibilities

- Design and build an RF oscillator in the 2.8–3.9 MHz region to drive a piezoelectric plate as the source of ultrasound. References [5,6] will show you how to do it.
- Design and build an experimental apparatus to produce the diffraction pattern of light from ultrasonically excited gratings in various liquids, such as water, castor oil, xylene, or toluene.
- Scan the diffraction pattern with a light sensor and compare the resulting intensities in various orders with theoretical predictions.
- Design and build an experimental apparatus to obtain the (Schlieren) striation pattern of a standing wave in a liquid on a distant screen, in order to

determine the sound wave velocity from the spatial period of the pattern and the geometry of the optics used.

- If the laser beam through the liquid cell covers a large enough area, the striation pattern may give clues to possible temperature and density variations in the liquid. By using suitable photodetectors or a digital camera, it might be possible to perform quantitative measurements of the intensity distribution in the striation pattern and make comparisons with theory.

Experimental suggestions

One needs a cell for the liquid with two transparent sides, a brass plate for reflecting the ultrasonic waves in the cell, an RF generator driving the piezoelectric quartz plate around 3 MHz, an oscilloscope, a frequency meter, several long- and short-focal lenses for beam broadening and imaging, a digital camera for recording the diffraction and the striation patterns, together with a rotating-telescope spectrometer for measuring diffraction angles accurately. For details on ultrasonic techniques the reader is referred to the literature [7,8].

One possible arrangement for observing the diffraction pattern is shown in Fig. 34.2 with approximate dimensions.

Direct observation of the ultrasonic grating, rather than just the resulting diffraction pattern of light scattered from it, may be made by using the experimental arrangement shown in Fig. 34.3, in which standing wave fronts are projected on a screen, as shown in Fig. 34.4.

Light waves exiting the liquid cell pass near the focal point of lens L_2 and then cast a "shadow" of the wavefronts on the screen. Typical dimensions are $f = 5\,\text{cm}$, $a = 1\,\text{m}$. The distance b needs very careful adjustment in order to be able to produce a sharp image on the screen or camera. The period p of the pattern observed on the screen, together with measured values of the distances f and g, and of the excitation frequency v, enables one to calculate the sound velocity c. A careful discussion of the shadow pattern in terms of ray optics is given in [4] (on which this project is based).

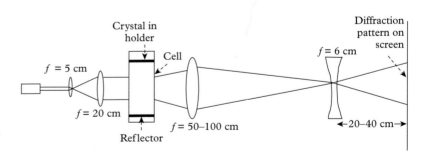

Figure 34.2 *Arrangement for diffraction pattern.*

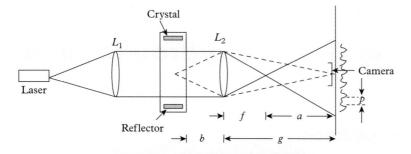

Figure 34.3 *Direct observation of grating.*

Figure 34.4 *Ultrasonic diffraction grating.*

REFERENCES

1. P. Debye and F.W. Sears, On the scattering of light by supersonic waves, *Proc. Nat. Acad. Sci. (US)* **18**, 409 (1932).
2. Ch. Bachem, E. Hiedemann, and H.R. Asbach, New methods for direct visualisation of ultrasonic waves and for the measurement of ultrasonic velocity, *Nature* **133**, 176 (1934).
3. B.D. Cook and E. Hiedemann, Diffraction of light by ultrasonic waves of various standing wave ratios, *J. Acoust. Soc. Amer.* **33**, 945 (1961).
4. P. Kang and F.C. Young, Diffraction of laser light by ultrasound in liquid, *Am. J. Phys.* **40**, 697 (1972).
5. Building the Pierce rf oscillator, video available on line at <http://www.youtube.com/watch?v=V1bfrJjSzw>.
6. H. Lythall, Basic RF oscillator, available online at <http://www.sm0vpo.com/blocks/osc7m00.htm>.
7. R.T. Beyer and S.V. Letcher, *Physical Ultrasonics* (Academic Press, New York, 1969), p. 221.
8. B. Carlin, *Ultrasonics* (McGraw-Hill, New York, 1960), p. 144.

35 The circular hydraulic jump

Introduction 188
Theoretical ideas 188
Goals and possibilities 189
Experimental suggestions 189

Introduction

Continuum mechanics has a rather small place allocated to it in the undergraduate education of physics majors. In hydrodynamics attention is usually restricted to Poiseuille flow and the Bernoulli equation. Yet the subject is rich in unusual phenomena which afford opportunities for open-ended experimental investigations and modeling.

Studying the circular hydraulic jump can proceed at many levels of competence and sophistication, as described in Refs [1–3]. We propose here a program which serves as an easy entry into the phenomenon, with a good chance of successful observation, and provides opportunities for the researcher to refine and extend the measurements, and to fit the results with the various models available in the literature.

The idea is to observe the impact of a vertical stream of liquid on a fixed horizontal plate. After impact the liquid spreads radially in a thin layer, but at some radius R from the place of impact its thickness increases suddenly – this is called the hydraulic jump. Understanding how this radius behaves as a function of the relevant physical parameters is the goal of the project.

Theoretical ideas

For a given nozzle diameter, the relevant parameters governing the phenomenon are the kinematic viscosity ν of the liquid, the flow rate q of the jet, and the height of release d from the faucet. In addition, boundary conditions will affect the radius, or wipe out the effect completely if, for instance, a boundary-free plate is replaced by one with, say, circular walls (a bowl).

Because of the complexity of a full hydrodynamical treatment [4,5], we restrict our considerations to scaling laws. For a given nozzle size, the radius R of the hydraulic jump is expected to follow the following power-law behavior:

$$R \sim q^\alpha d^\beta \nu^\gamma. \tag{35.1}$$

Establishing such a scaling law can follow several routes. The simplest ones are based on the model of Godwin [1] concerning the role of the viscous boundary layer. At a small radial distance r from the place of impact the liquid is assumed to spread outwards in laminar, virtually non-viscous flow, with a velocity $V_0 = \sqrt{(2gd)}$, except in a boundary layer near the surface of the plate of thickness $\sqrt{(\nu r/V_0)}$, in which the flow is viscous and laminar. The assumption is that the jump occurs when the boundary layer reaches the total height of the liquid film (fully developed boundary layer). Combining this assumption with a plausible parabolic velocity profile from the plate up in the vertical direction, the following values are found [3] for the scaling parameters:

$$\alpha = 2/3, \ \beta = -1/6, \ \gamma = -1/3. \tag{35.2}$$

These values are only a point of departure for the experimental investigation. More complex analytical treatments give slightly different results, for which we refer to the literature [2,4,5]. A theoretical discussion at an undergraduate level is given in [3], which also forms the experimental basis for this project.

Goals and possibilities

- Measure the jump radius R as function of the flow rate q for a particular liquid.
- Determine the dependence of R on the height d of the nozzle above the plate.
- For given height observe how R scales with viscosity ν. Since identical flow rates are difficult to achieve for various viscosities, the scaling of R with flow rate, obtained previously, will have to be used in this part of the project.
- Use video techniques to study the shape and height of the jump.
- Use differently textured plates to study their influence on the phenomenon.
- Tilt the impact plate to study the effect of this angle.
- Study the effect of boundary walls by using bowls of various diameters. Clearly, there are possibilities here for devoting anything from several days to many weeks to such a study.

Experimental suggestions

We follow the suggestions outlined in [3] and amplify them. An experimental setup for studying the influence of the flow rate, drop height, and boundary conditions is shown in Fig. 35.1. Water is the liquid for this part. The plate on which the liquid jet falls is 30×30 cm, thick glass or plexiglass, firmly and rigidly fixed to a support frame to prevent vibrations. Under the plate insert a sheet of polar or millimeter paper, protected from wetting by a thin glass plate, for measuring the radius of the jump. To protect the paper, seal the pair of plates all round with

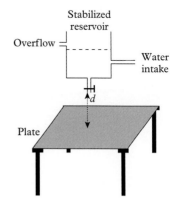

Figure 35.1 *Setup for observing the hydraulic jump.*

silicone or a suitable glue. Water issues from a glass tube, say of diameter 6 mm, attached to a plastic pipe provided with a clip which can regulate the flow rate, care being taken to obtain a steady stream without breakup into droplets. Another way to vary flow rates would be to use a series of outflow tubes varying, in 1 mm steps, between 3 and 10 mm in diameter. Typical practical drop heights are in the range 5–70 cm. The range of flow rates will depend on the overflow arrangement and the available pipes and clips; 10–70 ml/s is useful for the plate size recommended. Flow rates could be measured by suspending the vessel from a force sensor sampling its weight as a function of time.

Different-sized flat plates could be investigated in order to see their effect on the radius; also, try a glass beaker instead of a plate, in order to see the effect of vertical boundaries.

Constant flow rates can be achieved by using a stabilized container in which the water level is kept constant: it has an outflow opening at the bottom, an overflow opening assuring the level height (like the overflow protection in some water closets), and an intermediate opening between the two where the supply is replenished from the water mains. A cylindrical plastic or vessel some 3–5 l in volume is suitable. A different arrangement that would guarantee a constant pressure head, eliminating the need for the overflow opening in the previous method, would be the insertion of a tube into the vessel reaching a fixed height inside, its other end jutting out from the vessel. Any liquid beyond it simply drains out through its other end connected to a hose draining into a sink.

The setup for investigating the effect of viscosity is shown in Fig. 35.2. This part is harder and requires more care and planning. Because one deals with smaller quantities of dearer liquids, these have to be collected and reused. The input through the intermediate opening into the stabilized container is from another vessel containing the same liquid, rather than the water mains. The reservoir is filled to a certain level with liquids of various viscosity, flow rates are adjusted that are suitable for each liquid, all measurements should be performed from the same drop height. After each liquid the apparatus must be cleaned before proceeding with the next one. A suitable range of viscosities for five liquids would be the following: SAE 10 oil, SAE 30 oil, water, alcohol, ammonia. The temperature of the liquid and the apparatus must be known and kept as near constant as possible. The values for the viscosities are available from handbooks, or could be measured by the method of Stokes (falling ball bearing in a tall jar containing the liquid).

It is important to ascertain in each measurement that the flow rate q is a constant. In order to establish a reliable scaling law, q must be varied as much as practicable. It is vital to repeat the measurement of R for each q a number of times, in order to increase reliability. Figure 35.3, reproduced from [3], shows the dependence of the radius of the hydraulic jump on the flow rate q for different falling heights d.

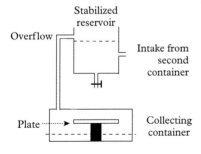

Figure 35.2 *Setup for observing the effect of viscosity.*

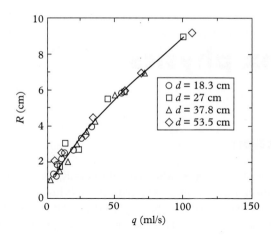

Figure 35.3 *Dependence of radius on flow rate.*

REFERENCES

1. R.V. Godwin, The hydraulic jump, *Am. J. Phys.* **61**, 829–32 (1993).
2. T. Bohr et al., Shallow water approach to the circular hydraulic jump, *J. Fluid Mech.* **254**, 635–48 (1993).
3. Y. Brechet and Z. Neda, On the circular hydraulic jump, *Am. J. Phys.* **67**, 723–31 (1999).
4. E.J. Watson, The spread of a liquid jet over a horizontal plane, *J. Fluid Mech.* **20**, 481 (1964).
5. I. Tani, Water jump in the boundary layer, *J. Phys. Soc. Jap.* **4**, 212 (1949).

36 Vortex physics

Introduction 192
Theoretical ideas 192
Goals and possibilities 194
Experimental suggestions 194

Introduction

Stirred coffee, bathtub vortices, whirlpools, dust devils, and tornadoes are some of the phenomena that are often observed. Yet students are never confronted with a laboratory experience of vortices, even though at least some of the tools for quantitative measurements are readily available, as outlined in the following project description.

The experimental apparatus used for this purpose is a magnetic stirrer found in chemistry laboratories, used for mixing materials dissolved in liquids. When a magnet is rotated at adjustable speeds at the bottom of a container containing a liquid, a large funnel-shaped vortex develops above it. Figure 36.1 shows a sketch of the physical parameters of the container, the liquid, and the vortex: $2a$ and d are the length and width, respectively, of the stirrer bar, R the radius of the cylindrical vessel, and H the height of still water. During stirring: H' is the displaced water height, h the distance between the funnel's lowest point and the bottom of the container, $\Delta h = H' - h$ and b the funnel's depth and half-width, respectively. The project is based on the experimental and theoretical work of [1].

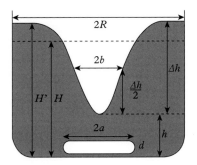

Figure 36.1 *Vortex parameters.*

Theoretical ideas

Vortex models

Several texts and internet sites provide an exposition of vortex hydrodynamics [1–7]. These propose theoretical formulae for the various components of the velocity field distribution within the vortex. Understanding them will be especially relevant if the laboratory for the project has access to research-grade particle image velocimetry (PIV) equipment, enabling one to map the liquid velocity field in the vortex. But even without this facility much physics can be learned and experienced.

For an infinite fluid the two simplest models for the velocity field are those of Rankine and Burgers. They are expressed in terms of the radial, tangential, and axial velocity components v_r, v_t, and v_z in cylindrical polar coordinates.

The Rankine vortex

See Refs [5,6]. Here, the vorticity is uniformly distributed in a cylinder of radius c, its axis being the z-axis. The radial and axial components of the velocity vanish and the tangential component has a discontinuity at $r = c$:

$$v_t = Kr/c^2 \text{ for } r \leq c, \text{ and } v_t = K/r \text{ for } r > c. \qquad (36.1)$$

Thus the liquid rotates as a rigid body for $r < c$. The parameter K is proportional to the circulation of the flow.

The Burgers vortex

In this model [7] the kinematic viscosity v is taken into account, the discontinuity in v_t is smoothed out, none of the velocity components vanish, and there is an axial inflow and an outflow maintaining the steady rotation:

$$v_r = -2vr/c^2, \; v_t = K[1 - \exp(-r^2/c^2)]/r, \; v_z = 4vz/c^2. \qquad (36.2)$$

In this model c is some effective radius within which the tangential flow is approximately a rigid-body rotation.

Dimensional analysis

In order to get an idea of the dependence of the funnel depth Δh and the half-width b on the geometrical and kinematic parameters, consider the Reynolds and Froude numbers, Re and Fr, defined below. When their values are small, viscosity and gravity effects are important, respectively. For the experimental system at hand the rotational frequency Ω of the stirrer bar governs the shape and size of the funnel and must therefore appear in both numbers:

$$Re = \Omega a^2/v, \text{ and } Fr = \Omega a/(gR)^{1/2}. \qquad (36.3)$$

Actually, it is more convenient to work with a modified Froude number $Fr' = Fr(d/a)$. For typical values, $R = 10\,\text{cm}$, $a = 2.5\,\text{cm}$, $d = 1\,\text{cm}$, $v = 10^{-2}\,\text{cm}^2/\text{s}$, $\Omega = 50\,\text{s}^{-1}$, one has $Re = 3 \times 10^4$ and $Fr' = 0.5$. This means that gravity plays an important role, but not viscosity.

Experimentally it is found that the funnel depth Δh is roughly proportional to both Ω^2 and a^2, and for $R > 0.4H$ scales as $R^{-1/2}$. Therefore, as a first hypothesis, we assume that $\Delta h/d$ depends linearly on Re and Fr', and is some unknown function of the remaining dimensionless parameters a/d, a/R, and H/R:

$$\Delta h/d = Re\, Fr'\, f(a/d, a/R, H/R). \qquad (36.4)$$

With the experimental dependencies of Δh ($\propto \Omega^2$, a^2) this may be simplified to

$$\frac{\Delta h}{d} = \frac{\Omega^2 a^2 d}{v(gR)^{1/2}} \Phi\left(\frac{H}{R}\right). \qquad (36.5)$$

Here, $\Phi(H/R)$ is some scaling function which could be fitted from the measured values of $\Delta h/\Omega^2$.

Goals and possibilities

- Measure the rotation frequency of the stirring bar using a stroboscope of adjustable frequency f.
- Measure the parameters of the funnel, in particular funnel depth as a function of Ω, for different values of H, R, A, and d.
- Use PIV in order to study the velocity field in a horizontal plane cutting through the funnel, as defined by an expanded laser beam.
- By spreading small light plastic beads in the liquid and tracing their movement as a function of time with a video camera, the central portion of the funnel vortex may be studied.
- A qualitative picture of the flow may be gained by injecting a dye, say food coloring, into the top of the funnel and observing its spreading visually or by photographic means.
- Small, very light objects of different shapes, sizes, and densities may be introduced into the center of the vortex, in order to gain some idea of the local vorticity by measuring their angular velocities.
- Devise some means to constrain the stirrer bar's movement in such a way that it should rotate around a fixed vertical axis.

Experimental suggestions

Obtain a series of cylindrical glass containers and magnetic stirrer bars of different dimensions: R ranging from 4 to 20 cm, H between 12 and 30 cm, and a from 1.5 to 4 cm.

Measure H' and h using a ruler. Measure the half-width of the funnel b on the back side of the cylinder. Take into account the curvature of the cylindrical container.

In the PIV method, photographing the positions of fine tracer particles at two times separated by 10^{-2}s enables one to determine the displacement and velocity of the particles in a horizontal plane.

The bar's motion will appear to be frozen whenever its frequency Ω equals an integer multiple of half a period between flashes of the stroboscope. If the corresponding frequency of the stroboscope is f_n then this happens for $\Omega = n\pi f_n$. Figure 36.2(a) shows the situation for $\Omega = \pi f_1$, the frozen bar and its mirror image in the container, while Fig. 36.2(b) shows the bar when it rotates at right angles between two flashes of the stroboscope, which happens when at $f_{1/2}$.

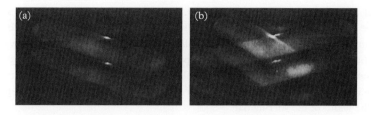

Figure 36.2 *Frozen bar: (a) f_1, (b) $f_{1/2}$.*

Use a cylindrical lens in order to obtain a horizontal sheet of laser light with which to define the bottom of the vortex and measure Δh.

The magnetic stirrer will have a sideways motion while rotating, and some means could be devised to minimize this. A possible solution is shown in Fig. 36.3: fix two identical stirrers into a tube with a hole in the middle through which a thin rod is fixed and held in such a way that the stirrers rotate about it as a fixed axis.

Use low density polyethylene beads of ≈ 1 mm diameter and monitor their motion by video as they are injected into the funnel. They slide down the surface of the funnel even though their density is less than 1 g/cm³, and will fluctuate below the funnel in an up–down motion, which looks as though it is chaotic, as shown in the traces of their motion in Fig. 36.4, taken over different time periods.

By measuring the velocity of rising of the beads in still water, an idea can be gained of the velocity of the downward jet needed to counter their rising and provide some vertical stability.

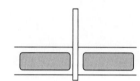

Figure 36.3 *Two stirrers.*

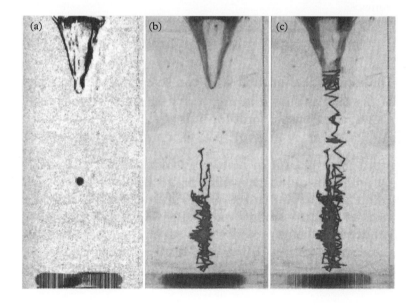

Figure 36.4 *Bead motion traces over time.*

If one injects a small amount of dye into the middle of the funnel, it will form some sort of cylindrical dye curtain around the axis of the vortex. Its stability, lifetime, and dimensions (close to the half-width b) could be studied as one varies the parameters of the system, and compared with the lateral extent of the paths of the plastic beads. The stability of such an envelope testifies to a lack of radial velocity in this region around the axis of rotation, of a width which must depend on the width of the stirring bar.

A qualitative picture of the pattern generated by the rotating stirrer is shown in Fig. 36.5. It is characterized by the streamlines (the thin continuous curves) and by a strong downward jet shown with the axial bold straight line. The two vertical dotted lines on either side of it define the approximate region of formation of the dye envelope.

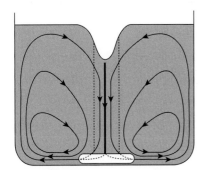

Figure 36.5 *Streamlines and downward jet.*

REFERENCES

1. L.D. Landau and E.M. Lifshits, *Fluid Mechanics* (Pergamon, Oxford, 1987).
2. H.J. Lugt, *Vortex Flow in Nature and Technology* (Wiley, New York, 1983).
3. B. Lautrup, *Continuum Physics: Exotic and Everyday Phenomena in the Macroscopic World* (CRC Press, Boca Raton, FL, 2011), ch. 21.
4. G. Halasz, B. Gyure, I.M. Janosi, K.G. Szabo, and T. Tel, Vortex flow generated by a magnetic stirrer, *Am. J. Phys.* **75**, 1092 (2007).
5. Wikipedia, Rankine vortex, online at <http://en.wikipedia.org/wiki/Rankine_vortex>.
6. Rankine vortex: A simple hurricane model, online at <http://demonstrations.wolfram.com/RankineVortexASimpleHurricaneModel/> and <http://www.youtube.com/watch?v=e4l_oV-S82I>.
7. P. Drazin and N. Riley, *The Navier–Stokes Equations: A Classification of Flows and Exact Solutions* (Cambridge University Press, Cambridge, 2006).

Plastic bottle oscillator

37

Introduction	197
Theoretical ideas	197
Goals and possibilities	198
Experimental suggestions	199

Introduction

The everyday phenomenon of emptying a bottle of liquid by turning it upside down, the "glug-glug" effect, is used in this project to study the periodic switching between the outflow of liquid and the inward flow of air into the bottle. The phenomenon is surprisingly rich in physics. The diameter of the opening determines which of the three regimes of *no flow*, *oscillatory flow*, or *counter flow* will be observed. If two such bottles are coupled by pipes, we have coupled oscillators which will oscillate in phase or antiphase, depending on whether they are joined by a straight pipe or one in the shape of a U-tube. The project described here is based on [1].

Theoretical ideas

As soon as the bottle is turned upside down water moves down due to gravity; the gauge pressure inside the bottle gradually decreases and this will retard the flow. When the inertia of the downflow is not large enough to counteract this pressure difference, the downflow will stop and a steady state of no flow will be reached. This happens for small-diameter pipes draining the bottle.

For oscillatory flow the damping effect of the pipe is less than before, and transient flow occurs for a while, after the total pressure attains equilibrium. As a result of damping the downflow ceases, the pressure difference is reversed in sign, generating instead an air flow inward. This upward air flow is also arrested and the cycle begins afresh, as shown in Fig. 37.1.

When the pipe diameter is large enough, simultaneous water and air flows are possible and there is no oscillation.

When the bottles are connected by a straight hollow pipe, as illustrated in Fig. 37.2, the air pressure in both bottles is the same. This must be due to the different time scales of the relaxation time of the pressure difference and the period of oscillation. The tendency for the pressure difference between the bottles to stay small is reflected in the in-phase oscillation of water and air flow in the two systems.

Connecting the bottles with a U-tube containing water will change the phase. The water in the tube separates the air spaces of the bottles. If there is a phase difference between the bottles, the water in the interconnecting tube moves to

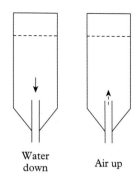

Figure 37.1 *Single bottle.*

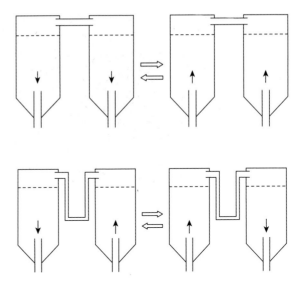

Figure 37.2 *Synchronized in-phase oscillation.*

Figure 37.3 *Oscillations in anti-phase.*

reduce the pressure difference between the two bottles. The interaction between the bottles has the opposite sign to that in the case of the hollow tube and results in antiphase oscillations of the flows, as shown in Fig. 37.3.

The phenomenon of periodic switching between two different states, outflow of water and inflow of air, is reminiscent of other oscillatory systems in which at some value of the coupling parameter (here represented by hollow and water-filled connecting tubes) a bifurcation occurs. Some examples of such systems are saline oscillators, coupled metronomes, the pouring of water into a cup, and the synchronous flashing of fireflies [2–5].

Goals and possibilities

- Find the maximum diameter of the drain pipe for which no flow occurs.
- Measure the period of oscillation, water outflow alternating with air inflow, for a drain pipe of a particular diameter.
- Measure the period of oscillation as a function of drain pipe diameter.
- Find the minimum diameter of the drain pipe for which the downflow of water and the upflow of air can occur simultaneously.
- Rotate the bottle at a variable angular velocity, whereby the "tornado-in-the-bottle" vortex effect will influence the draining time, periodicity, and oscillatory behavior.

- Connect the two bottles by an empty pipe above the maximum water level and investigate the synchronization of the periodic draining in both bottles.
- Connect the two bottles by a U-tube containing water and explore the change in phase of the oscillation of one bottle relative to the other.
- Set up and study a differential equation to model the dynamics of the bifurcation, containing the important parameters of the system: the density of the liquid, the compressibility of the air in the bottle, the elasticity of the bottle, the diameter of the draining pipe, gravity, and buoyancy.

Experimental suggestions

Two-liter plastic bottles are suitable, filled with, say, 1.8 l of water. Use drain pipes of length 100 mm and inner diameter ranging from 2 to 24 mm.

Dye the water with some ink in order to be able to photograph it easily as it emerges. The time-dependent changes in the outflow may be followed by filming them with a digital video camera focused just below the opening. The dyed water will clearly distinguish the alternating start and stop of the outflow.

Frame-by-frame analysis by a frame grabber enables one to draw a time series of the phenomenon.

Rotating the bottle while draining is tricky: one needs a method for holding the 2 kg bottle in some tight-fitting vise or envelope that is attached to a rod inserted into the chuck of a drill, as in Fig. 37.4.

Figure 37.5, reproduced from [1], shows sample measurements for three pipes, each of length 100 mm and inner diameters d of 6, 8, and 12 mm.

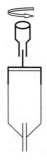

Figure 37.4 *Rotating the bottle.*

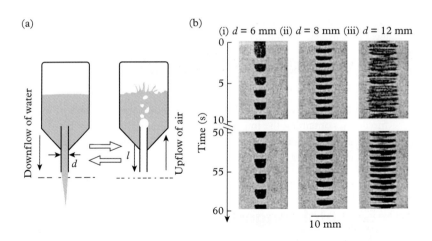

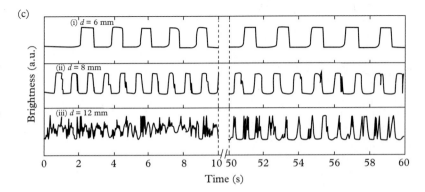

Figure 37.5 *(a) Experimental setup; the dashed line shows where horizontal images were aligned. (b) Spatio-temporal diagram of oscillation for three pipes of different diameters d: dark and light regions correspond to water downflow and air upflow, respectively. (c) Time trace of water flow, as represented by the brightness in the central part of the image: positive and negative values correspond to upflow and downflow, respectively.*

REFERENCES

1. M.I. Kohira, N. Magome, H. Kitahata, and K. Yoshikawa, Plastic bottle oscillator, *Am. J. Phys.* **75**, 893 (2007).
2. H.M. Smith, Synchronous flashing of fireflies, *Science* **82**, 151 (1935).
3. J. Pantaleone, Synchronization of metronomes, *Am. J. Phys.* **70**, 992 (2002).
4. K. Yoshikawa, N. Oyama, M. Shoji, and S. Nakata, Use of a saline oscillator as a simple non-linear dynamic system, *Am. J. Phys.* **59**, 137 (1991).
5. H. Kitahata, A. Yamada, and S. Nakata, Model bifurcation by pouring water into a cup, *J. Chem. Phys.* **119**, 4811 (2003).

Salt water oscillator

38

Introduction

When a small vessel having a narrow opening at its bottom and containing salt solution is submerged in a much larger vessel containing pure water, an interesting oscillatory phenomenon occurs: salt water begins to flow downwards, then stops and pure water flows upwards. The down flow and up flow alternate with a given period, depending on the parameters of the problem. This project, on the border between physics and physical chemistry, is concerned with the experimental investigation of the phenomenon and its theoretical understanding.

Introduction	201
Theoretical ideas	201
Goals and possibilities	202
Experimental suggestions	203

Theoretical ideas

The following are only qualitative ideas, and the reader will find a thorough theoretical treatment in [1]. See also [2].

The system is shown schematically in Fig. 38.1. A liquid B of higher density is in a small container with a narrow capillary C or a pinhole at its bottom. This opening is too narrow to allow two-way flow. The container is immersed in a large vessel containing liquid A of lower density.

Figure 38.1 shows two possible stationary states. In state 1 the capillary is full of salt solution, in state 2 it is full of pure water. The system oscillates between these two states as it approaches the equilibrium state of uniform concentration throughout.

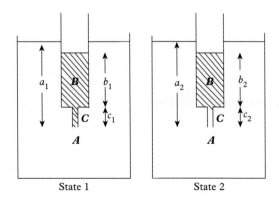

Figure 38.1 *Two possible stationary states.*

Let ρ_w, ρ_s be the densities of water and salt solution, respectively. Then, referring to Fig. 38.1, we have for the balance of pressures in state 1: $a_1 \rho_w = (b_1 + c)\rho_s$, while in state 2, $a_2 \rho_w = (b_2 + c)\rho_s$.

The surface area of A is much larger than that of B, so that a_1 and a_2 are almost equal. Therefore b_2 is larger than b_1, meaning that the fluid level in B is higher when the capillary is full of water than when it is full of the salt solution.

In the absence of fluctuations (and neglecting the very slow diffusion of salt across the entrance of the capillary) either of the two stationary states could persist. But if there are fluctuations, say a small amount of liquid from A were to enter the capillary, the pressure in B and C would not be able to balance that in A at the same level. Then water would be driven up from A to C with an ever-increasing driving force. Once the capillary became full of water instead of salt solution, flow would continue until the level in B reaches the value in stationary state 2. In the process ρ_s decreases.

Stationary state 2 is also unstable against fluctuations. Should a little salt solution enter the top of the capillary from B, the increased pressure will induce an accelerating flow until the capillary is full with salt solution. The flow from B will continue until state 1 is reached. In the process, ρ_w increases. Irrespective of the direction of flow, the free energy decreases and the final state is one of uniform composition everywhere. It is the fluctuations which are responsible for the change in the direction of flow and the fact that the system does not reach either of the stationary states monotonically.

Interestingly, oscillations in the flow direction also occur [3–5] if the capillary is replaced by just a pinhole at the bottom of B, in spite of the fact that only an insignificant difference in hydrostatic pressure separates the stationary states 1 and 2 in which the orifice is filled with salt solution or water, respectively. If the interface between the two liquids is just inside B, instability will drive water up from A instead of the salt solution escaping through the hole. This water influx results in an increase of the level in B, which continues until the hydrostatic pressure is large enough to reverse the flow, causing salt solution to flow down through the hole. It being denser it will sink into the container A, and the flow following it will continue until the level in B drops significantly.

Electrodes inserted into A and B will sense different potentials, related to different chloride concentrations. The changes in potential generated at the interface between dilute and concentrated salt solutions accompany the oscillations in the flow, and can therefore be used to monitor the latter.

Goals and possibilities

- Measure the period of pulsations of the flow for a given concentration of the salt solution.
- Vary the length and diameter of the capillary and see how this affects the period of oscillation.

- Try different salt concentrations.
- Try two liquids differing in density other than water and salt solution.
- Try the experiment with just a small hole at the bottom center of the salt water container.
- Conduct an interesting experiment, resulting in an alternating voltage battery, with two syringes filled with saturated copper sulphate pentahydrate ($CuSO_4 \cdot 5H_2O$) aqueous solution immersed in a reservoir of distilled water.

Experimental suggestions

For the salt solution use a 10 ml plastic syringe, modified at the bottom to fit a capillary of the chosen diameter (1.5 mm is a good starter) and length. Start with the level of the solution in the syringe nearly equal that of the pure water in the 400 ml beaker. Use 1 M, 3 M, and saturated NaCl solutions.

There are two possible ways to measure the period of oscillation of the up and down flows, one electrical the other optical. In the former, electrodes are inserted into the salt solution, as shown in Fig. 38.2, and the water and the oscillations in the potential difference are recorded as a function of time. Potential differences are of the order of tens of mV. Electrodes can be an Ag/AgCl pair. Since these are fairly expensive, an alternative method [1] would be to use as the pair of liquids saturated copper sulphate pentahydrate ($CuSO_4 \cdot 5H_2O$) aqueous solution on top and distilled water in the reservoir, in which case thin copper plates can serve as suitable electrodes.

The optical method consists of dying the salt solution with black ink, positioning a narrow laser beam just below the opening, and monitoring the transmitted light with a photodiode. In the down flow of salt water the beam is obstructed, whereas during the flow of pure water the beam is transmitted.

The setup for the alternating voltage battery [6] is shown in Fig. 38.3. Both syringes are filled with $CuSO_4 \cdot 5H_2O$ solution, each has a thin copper plate electrode.

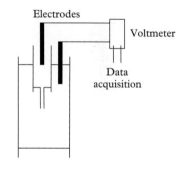

Figure 38.2 *Recording the oscillations.*

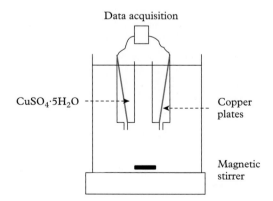

Figure 38.3 *Setup for AC voltage battery.*

To start the oscillations, extract a few drops of solution from the bottom of one of the syringes. This results in a phase shift between the two concentration-oscillating cells. A data acquisition system records the resulting difference between the EMFs of the cells.

REFERENCES

1. K. Yoshikawa, Oscillatory phenomenon chemistry – development of temporal order in molecular assembly, in *Dynamical Systems and Applications*, ed. N. Aoki, pp. 205–224 (World Scientific, Singapore, 1987).
2. R.M. Noyes, A simple explanation of the salt-water oscillator, *J. Chem. Edu.* **66**, 207 (1989).
3. K. Yoshikawa, S. Nakata, M. Yamanaka, and T. Waki, *J. Chem. Edu.* **66**, 205 (1989).
4. K. Yoshikawa, N. Oyama, M. Shoji, and S. Nakata, Amusement with a salt-water oscillator, *Am. J. Phys.* **59**, 137 (1991).
5. J. Walker, The salt fountain and other curiosities based on the different density of fluids, *Sci. Am.* **237**, 142 (1977).
6. R. Cervellati and R. Solda, An alternating voltage battery with two salt-water oscillators, *Am. J. Phys.* **69**, 543 (2001).

Part 5
Optics

39	Birefringence in cellulose tapes	207
40	Barrier penetration	212
41	Reflection and transmission of light	215
42	Polarization by transmission	221
43	Laser speckle	226
44	Light scattering from suspensions	232
45	Light intensity from a line source	236
46	Light interference in reflecting tubes	239

Part-5

Optics

Birefringence in cellulose tapes

39

Introduction	207
Theoretical ideas	207
Goals and possibilities	209
Experimental suggestions	210

Introduction

One can view through crossed polarizers beautiful color effects resulting from passing unpolarized white light through many layers of transparent adhesive tape [1]. Stresses during the manufacture of such tapes result in optical anisotropy, so that light having polarization components parallel and perpendicular to the tape length travel with different velocities, and therefore have different refractive indices. A stack of many layers of tape causes significant retardation of incident light. This project provides an opportunity to explore an attractive, easily accessible phenomenon quantitatively, by measuring the intensity of the light transmitted through the layers and its spectral composition. Before embarking on the project, study carefully a chapter on polarization in a textbook [1].

Theoretical ideas

Transmitted intensity

All of the following, which is merely a summary, is derived in the superb book by Crawford [1]. For a retarder of thickness s sandwiched between crossed polarizers, as in Fig. 39.1, we wish to relate the transmitted intensity I_t to the incident (unpolarized) light intensity I. In Fig. 39.1, the unit vectors a, b label the mutually perpendicular transmission axes of the polarizers. Uniaxial stress in materials often results in an axis of symmetry (sometimes called the optic axis of the retarder) with all perpendicular directions optically equivalent. So, in Fig. 39.1 we break the incident light into components polarized along this "extraordinary" axis (e) or polarized along an "ordinary" axis, o, perpendicular to e (x is the direction of propagation of the wave). Both the retarder and the polarizers are assumed to be perfect, with transmission coefficients equal to one (in the transmitting directions).

The intensity is the time average of the electric field,

$$I = \overline{|E|^2}/\mu_0 c. \tag{39.1}$$

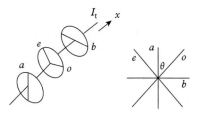

Figure 39.1 *Retarder between crossed polarizers.*

The electric field emerging from the first polarizer can be written in vector form:

$$\boldsymbol{E}_1 = E_1 \cos(2\pi[ft - (f/c)x])\boldsymbol{a}, \qquad (39.2)$$

where $E_1 = (\mu_0 cI)^{1/2}$. The wave exiting the retarder of thickness s, with $x = 0$ chosen at its entrance face, will be

$$\boldsymbol{E}_2 = E_1\{\cos\theta\cos(2\pi[ft - (n_0 f/c)s])\boldsymbol{a} + \sin\theta\cos(2\pi[ft - (n_e f/c)s])\boldsymbol{b}\}, \qquad (39.3)$$

where n_0, n_e are the two refractive indices, f is the frequency, and c the velocity of light. We see that the phase difference between the $\boldsymbol{a}$ and $\boldsymbol{b}$ components is $2\pi(n_0 - n_e)s/\lambda$. The second polarizer transmits only the $\boldsymbol{b}$ component, so that the wave exiting it will be

$$\boldsymbol{E}_t = E_1 \sin\theta\cos\theta\{\cos 2\pi[ft - (f/c)s] - \cos 2\pi[ft - (f/c) + R\lambda]\}\boldsymbol{b}, \qquad (39.4)$$

where $R = (n_0 - n_e)s$. Performing the time average in order to calculate the final intensity, one obtains

$$I_t/I = 0.25\sin^2 2\theta[1 - \cos(2\pi R/\lambda)]. \qquad (39.5)$$

The important parameter governing intensity maxima and minima is R/λ.

The Lyot filter

A Lyot filter [2] is a kind of optical filter, i.e. an optical device with a wavelength-dependent power transmission. It consists of a sequence of birefringent crystalline plates (e.g. of quartz) and polarizers. The birefringent axis of each crystal is oriented at 45° to the axis direction of the polarizers. The light propagating in a crystal can be considered as containing two different linear polarization components, which experience a different phase delay. The relative phase delay for the two polarization components depends on the wavelength. Therefore, the loss of optical power at the subsequent polarizer is wavelength dependent.

For a device with a single birefringent crystal, the power transmission versus optical frequency can be described with an approximately sinusoidal oscillation. (Chromatic dispersion causes some deviation from an exact sinusoidal oscillation.) By combining multiple crystals with different thicknesses, a sharper filter function can be realized. According to the Lyot design, the thickness of each crystal is half the thickness of the previous one – see Fig. 39.2. In this way, a small transmission bandwidth combined with a large period of the transmission peaks is possible – see Fig. 39.3.

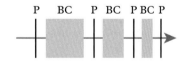

Figure 39.2 *A Lyot filter, consisting of a sequence of birefringent crystals (BC) and polarizers (P).*

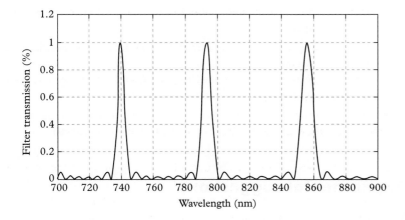

Figure 39.3 *Transmission function of a Lyot filter containing three quartz plates, with thickness values of 5, 2.5, and 1.25 mm.*

Goals and possibilities

- Do some hands-on experiments to illustrate and observe phase shift in polarization: prepare 16 layers of transparent Scotch tape stuck on a microscope slide (see the experimental suggestions section). Tape a polaroid, which has its axis at 45° to the tape axis, to one face of the package, together with a diffraction grating. Hold the package so as to look at a bright white light source and with the other hand hold another polaroid (a linear polarizer) at the exit side of the package and parallel to the first polaroid. Carrying out the following experiments [1] will help to understand polarization effects:
  1. Note the black bands, which have linear polarization absorbed by the linear polarizer. The bright region between two dark bands has relative phases ranging between zero and 2π.
  2. Rotate the analyzing polaroid by 90°, note and explain why the dark bands change into bright ones.
  3. Replace the polaroid by a circular polarizer with the output end toward your source. Note and explain the result.
  4. Put both linear and circular polarizer behind the stack, making sure you split the field of view between the two. By moving the tape stack look through the circular one first, then the linear one. Notice the movement of bands and interpret in terms of phase shifts. Then rotate the linear polarizer, thereby reversing the band motion, and interpret the effect.
- Use a grating spectrometer in order to find the intensity minima in (39.1) at suitable wavelengths.
- Find the retardation per layer of tape. From the data calculate the difference in refractive indices (the birefringence).

- Combine retarders at different angles and investigate the effects on the dark band in the spectrometer.
- Analyze the spectrum emerging from the retarder and the polarizers by means of a spectrophotometer, in order to measure accurately the transmission intensity as a function of wavelength. This will necessitate a revision of (39.5), based as it was on perfect polarizers and no absorption.
- Make your own Lyot filter (see next section) and repeat the above experiments, both the qualitative ones and with the spectrometer.

Experimental suggestions

Equipment needed: reel of clear cellophane Scotch tape, polaroid, spectrophotometer/grating spectrometer, diffraction grating, Polaroid sheets, circular polarizer, and a bright white light line source.

To obtain a good stack of Scotch tape layers, first place a small bead of mineral oil on a new microscope slide. Stick a length of transparent tape longer than the slide onto the slide and to the table. The oil should fill in any air gaps and ensures good optical contact between tape and slide. (The index of refraction of the oil is much closer to that of the tape and the slide than would be the case for an air bubble.) On top of that tape again place a bead of oil and stick a second piece of tape on the first. Carry on in this vein, alternating oil with tape, until you have the desired number of layers, 4, 8, 16, or 32, for the project. On the last layer put a bead of oil and cover with a second microscope slide. The sample is now ready for experiments.

Transmission experiments

White light falls on the sample mounted between crossed polarizer sheets. The light emerging from the second polarizer is allowed to fall on a grating spectrometer. The dark bands in the spectrum will correspond to the zeros of $(1 - \cos 2\pi R/\lambda)$, so that $R/\lambda_{min} = 1, 2, 3, \ldots$. One can then plot $1/\lambda_{min}$ versus integers, which should give a straight line of slope $1/R$.

Measure the thickness of the sample with a micrometer, in order to be able to calculate the difference in refractive indices. Figure 39.4, reproduced from [3], shows a typical set of data from 32 layers of tape, obtained with a grating spectrometer.

Lyot filter experiments

Make your Lyot filter from cellophane tape and pieces of Polaroid: you will need to make three retarders from 4, 8, and 16 layers; call them f, e, s, respectively. In addition, you'll need pieces of Polaroid polarizers (P for short) and a diffraction

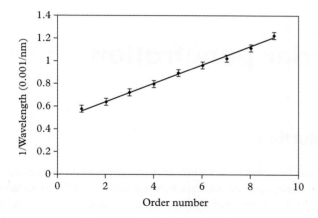

Figure 39.4 *A plot of reciprocal wavelength at which minima occur in the spectrum against integer order numbers. The inverse of the slope gives a retardation of 360±20 nm per layer of tape.*

grating (D for short). As in the hands-on experiments, the polarizer's axis is at 45° to the tape axis.

To get a feel for the filter make a D:P(45°):s:P(45°):e:P(45°) "sandwich." Then look at the line source. Now add another f:P(45°) sandwich to the first one: as successive filters are added the sidebands are wiped out. The bandwidth is given by the 16-layer stack. Adding a 32-layer stack would halve the bandwidth, but that would be asking too much from the transparency of the Scotch tape stack.

REFERENCES

1. F.S. Crawford, *Waves, Berkeley Physics Course* (McGraw-Hill, New York 1968), p. 440.
2. Lyot filters, *Encyclopedia of Laser Physics*, online at <http://www.rp-photonics.com/lyot_filters.html>.
3. S.D. Cloud, Birefringence experiments for the introductory physics course, *Am. J. Phys.* **41**, 1184 (1973).

40 Barrier penetration

Introduction 212
Theoretical ideas 212
Goals and possibilities 213
Experimental suggestions 213

Introduction

Total internal reflection occurs when electromagnetic waves are incident, beyond the critical angle, upon an interface between media of differing refractive indices, from the dense side. Nevertheless, a non-propagating solution to Maxwell's equation does "slosh" slightly into and out of the less-dense region; this *evanescent* wave decays exponentially to near zero within a distance of the order of a wavelength (as the angle of incidence approaches the critical angle, this penetration distance extends further and further into the less-dense medium, but for larger angles of incidence there is no detectable energy coupled into the less-dense region). So, when total internal reflection occurs, the incident energy is completely reflected into the dense medium.

On the other hand, if the spatial extent of the less-dense region is less than the penetration depth of the evanscent wave, it may be possible to couple energy to the other side: for example, suppose that a thin, low-index layer separates two identical dielectrics; the evanescent wave will penetrate the "forbidden gap" (i.e., the low-index layer) and the transmitted portion may be detected (again, provided the width of the gap is of the order of the wavelength). The first part of this project is devoted to measuring the transmission coefficient for light waves passing through a variable "forbidden gap" of air between two glass prisms, as a function of the gap width. The second part is devoted to exploration of such evanescent waves for microwaves.

Theoretical ideas

Consider the passage of a plane light wave through two 45°–90°–45° prisms, as shown in Fig. 40.1.

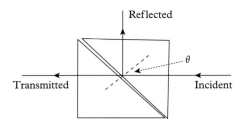

Figure 40.1 *Transmission through two prisms.*

The angle θ between the incident ray and the normal to the surface where total internal reflection occurs is taken to be greater than the critical angle. The transmission coefficient T for the transmitted light, the ratio of intensities, is calculated by solving Maxwell's equations [1] using appropriate boundary conditions at the two interfaces. Let d be the width of the air gap, θ the angle of incidence on the gap, ε and μ the dielectric constant and permittivity of glass, respectively. The transmission coefficient T is found from the ratio of Poynting vectors. For the electric field $\boldsymbol{E}$ parallel to the plane of incidence, T is given by

$$T = A/[A + 4\sinh^2(\kappa d)], \tag{40.1}$$

where A, κ are given by

$$A = 16[k\kappa/(k^2 + \kappa^2)]^2,$$
$$k = (2\pi/\lambda)[\mu/\lambda]^{1/2}\cos\theta,$$
$$\kappa = (2\pi/\lambda)[n^2\sin^2\theta - 1]^{1/2},$$
$$n = [\mu\varepsilon]^{1/2}.$$

For $\boldsymbol{E}$ perpendicular to the plane of incidence, ε and μ are interchanged in the expression for k. Equation (40.1) is suggestive: it is in the form of the transmission coefficient for a particle tunneling through a rectangular barrier in quantum mechanics. T vanishes when $d = 0$, and when θ is equal to the critical angle. For $d \geq \lambda$, it is an exponential.

Goals and possibilities

- Build an optical setup in which both the angle of incidence and the width of the air gap can be varied continuously.
- Measure the transmitted intensity as a function of the air gap width.
- Consider transmission measurements for various wavelengths.

Experimental suggestions

Two 45° crown glass prisms with 2.5 cm-long diagonal faces are aligned, as in Fig. 40.1. A 10 μm-thick mica sheet inserted at one end of the diagonal interface of the two prisms serves as a wedge, so that the air gap varies continuously from zero to 10 μm along 2.5 cm. The useful part of the gap is furthest from the wedge, near the contact point of the two faces [2].

Design a method of clamping the two prisms on a rotating turntable, in order to be able to vary the angle θ between the incident ray and the prism surface where total internal reflection occurs. The prisms should be encased in brackets, one fixed and the other micrometer controlled, enabling one to press the diagonal

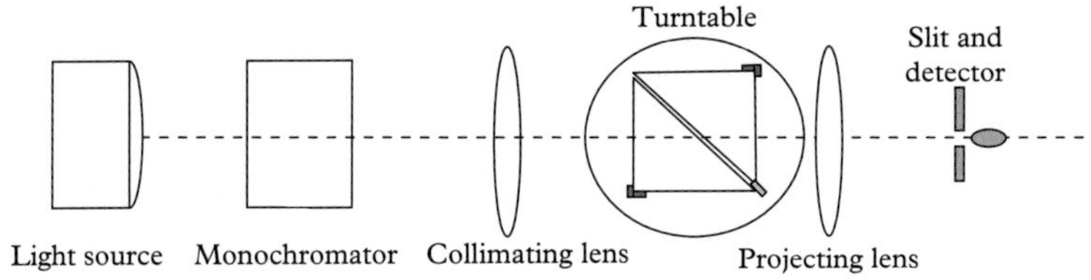

Figure 40.2 *Setup for measuring transmitted intensity.*

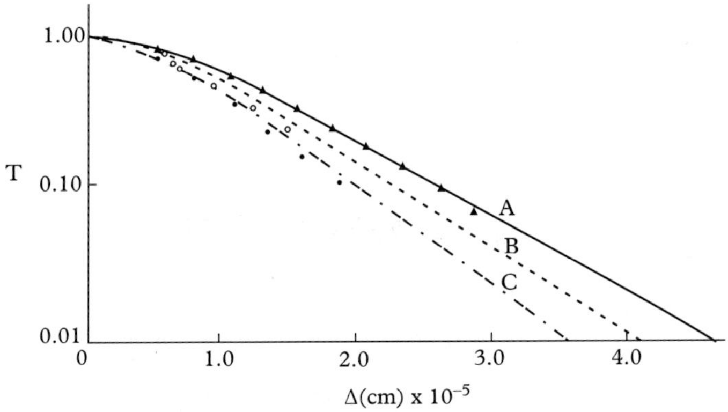

Figure 40.3 *Transmittance as a function of the air gap Δ. Lines A, B, C are plots of (40.1) for 575, 507, and 450 nm, respectively. Triangles, circles and full circles are data for $\theta = 55°$.*

faces together in a controlled way to just the right amount (full light transmission at the contact end), without distorting the prism surfaces. The interference fringes from the wedge surface will be a good indicator of the right position.

An intense xenon or other lamp, combined with a monochromator and collimating lens, can serve as light source, allowing experimentation with a number of wavelengths. Alternatively, single-wavelength measurements can be made using a laser. The light is incident at an angle $\theta > \theta_{crit}$, achieved by suitably rotating the system, and restricting the tunneling to a small region near the contact edge of the prisms.

A short-focal-length cylindrical lens projects an image of the contact region to a screen several meters away, enabling one to see the transmitted light.

The intensity of the transmitted light may be measured by a movable slit–photodetector combination, which can thus track the light emitted by transmission at various points of the gap. A suggested setup is shown in Fig. 40.2.

Figure 40.3, reproduced from [2], shows transmittance as a function of the air gap Δ.

....

REFERENCES

1. D.D. Coon, Counting photons in the optical barrier penetration experiment, *Am. J. Phys.* **34**, 240 (1966).
2. J.C. Castro, Frustrated total internal reflection, *Am. J. Phys.* **43**, 107 (1975).

Reflection and transmission of light

41

Introduction 215
Theoretical ideas 215
Goals and possibilities 217
Experimental suggestions 218

Introduction

The Fresnel equations, together with appropriate boundary conditions and the Poynting expression for the intensity, determine the percentage of reflected and transmitted light energy between two media differing in refractive indices, for all angles of incidence and polarizations. Although this is standard material in an electromagnetic theory course, not many students actually do quantitative experiments on this topic. The present project gives an opportunity in two different experimental setups to measure the reflected and transmitted intensities – one near the critical angle for various polarizations, the other for all angles and a particular polarization. Although there are no surprises, much physics can be learnt, not the least of which is the need to be in complete control of all factors influencing the measurements.

Theoretical ideas

The first experiment considers a monochromatic beam of light incident at angle i_1 on one face of a 45°–90°–45° prism, exiting on the diagonal face at angle r_2, and at angle i_1 on the lower face – see Fig. 41.1.

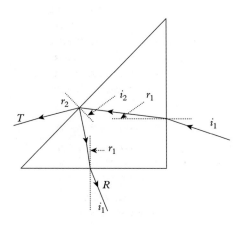

Figure 41.1 *Geometry of light paths through prism.*

216 *Physics Project Lab*

Of interest are the fractions of energy in beams R and T, calculated by applying the above-mentioned method at the air–glass interfaces [1]. Resolution into polarizations perpendicular to and parallel with the plane of incidence gives the following relations for the R and T coefficients:

$$R_\perp^2 = \frac{4n \cos i_1 \cos r_1 (n \cos i_2 - \cos r_2)}{(\cos i_1 + n \cos r_1)^2 (n \cos i_2 + \cos r_2)}, \tag{41.1}$$

$$R_\parallel^2 = \frac{4n \cos i_1 \cos r_1 (\cos i_2 - n \cos r_2)}{(\cos i_1 + n \cos r_1)^2 (\cos i_2 + n \cos r_2)}, \tag{41.2}$$

$$T_\perp = \frac{16n \cos i_1 \cos i_2 \cos r_1 \cos r_2}{(\cos i_1 + n \cos r_1)^2 (n \cos i_2 + \cos r_2)}, \tag{41.3}$$

$$T_\parallel = \frac{16n \cos i_1 \cos i_2 \cos r_1 \cos r_2}{(n \cos i_1 + \cos r_1)^2 (\cos i_2 + n \cos r_2)}. \tag{41.4}$$

In the second experiment a spherical flask is half-filled with water and a laser beam is aimed at its center (see below for details), as in Fig. 41.2. The dominant light fluxes emerging are R_1 and T_1. Their intensities are determined by reflections and transmissions at the liquid–air, liquid–glass, and air–glass interfaces, and by the exponential attenuation of light traversing the radial distance with absorption coefficient α. At the air–glass (ag) and liquid–glass (lg) boundaries incidence

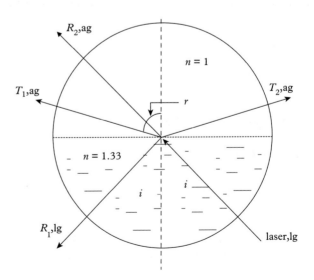

Figure 41.2 *Light paths and intensities through a spherical flask.*

is normal, for which the expressions for reflectance and transmittance coefficients R, T are particularly simple. If n_1 and n_2 are the refractive indices, we obtain

$$R = \left(\frac{n_2 - n_1}{n_2 + n_1}\right)^2, \qquad (41.5)$$

$$T = \frac{4 n_1 n_2}{(n_2 + n_1)^2}. \qquad (41.6)$$

At the glass–liquid boundary R is small, since the difference in refractive indices is small. Adjacent interfaces have resultant coefficients which are products of the individual ones. At the air–liquid (al) interface incidence is oblique and R, T depend on the polarization:

$$R_\perp = \frac{\sin^2(i - r)}{\sin^2(i + r)}, \qquad T_\perp = \frac{4 \sin i \cos i \sin r \cos r}{\sin^2(i + r)}, \qquad (41.7)$$

$$R_\parallel = \frac{\tan^2(i - r)}{\tan^2(i + r)}, \qquad T_\parallel = \frac{4 \sin i \cos i \sin r \cos r}{\sin^2(i + r)\cos^2(i - r)}. \qquad (41.8)$$

We have, therefore, for beams R_1 and T_1, using (41.7) for parallel polarization,

$$R_1 = e^{-2\alpha r} T_{ag}^2 T_{lg}^2 R, \qquad T_1 = e^{-\alpha r} T_{ag}^3 T_{lg} T_\parallel. \qquad (41.9)$$

For perpendicular polarization one uses (41.8) instead. Unpolarized light would be described by averages of both components.

Beam T_2 results from back reflections of R_1 and T_1 at the glass envelope, and if the flask is exactly half full it will be symmetrical with respect to T_1. The expression for its relative intensity accounts for multiple reflections and absorptions:

$$T_2 e^{-\alpha r} T_{ag}^3 T_{lg} R_{ag} TR(1 + T_{ag}^2) + e^{-3\alpha r} T_{ag}^3 T_{lg} TR(R_{lg} + T_{lg}^2 R_{ag}), \qquad (41.10)$$

where T and R are taken from (41.7) or (41.8).

Goals and possibilities

Prism apparatus

- Measure the refractive index of the glass prism by the minimum deviation method.
- Use the refractive index so measured to calculate the reflection and transmission coefficients in both experimental setups and plot them as a function of the angle of incidence for both polarizations. Particular care is to be taken

in using formulae (41.1)–(41.4) when i_2 exceeds the critical angle: in this case $R = 1$ and $T = 0$ at the diagonal boundary.

- Measure the angle-dependent reflection and transmission coefficients for beams R and T for different polarizations by using a polarizer in the path of the incident beam. Compare with the theoretical curves, suitably normalized.
- From the intensity measurements determine which i_1 will give the critical angle i_2 and compare with the theoretical value calculated using the refractive index already measured.

Spherical flask apparatus

- Determine the absorption coefficient in water by measuring intensities with flask full and empty.
- Determine the refractive index of the water in the flask by measuring the vertical projections of the R_1 and T_1 beams and using the geometry of the flask.
- Calibrate the photodetector and measure the intensities when a polarizer is put into the path of the laser beam to obtain parallel or perpendicular polarization. Compare with the theoretical expressions, (41.7)–(41.10).
- Find the angle of the polarizer with respect to the beam which gives maximum and minimum intensities; finding the "Brewster angle" can be used to estimate the index of refraction.

Experimental suggestions

Prism setup

See Ref. [1]. The experiment is best performed in a dark room. For a detector use a silicon solar cell, connected to a microammeter. The cell is in a black box enclosure, with suitable apertures, in order to reduce stray ambient radiation. Calibrate the cell by using a series of neutral density filters to vary the intensity in a controlled manner. For the source use a standard He–Ne laser.

The glass prism is set on the rotating table of a spectrometer. Clean the prism faces thoroughly to avoid scattering by dust particles. The normal to the prism face near the laser, and hence the reference line for incident angles, is determined by turning the prism until the laser beam is reflected upon itself.

For each angle of incidence repeat the measurement of the laser beam intensity, since it may fluctuate with time.

Reflection and transmission of light   219

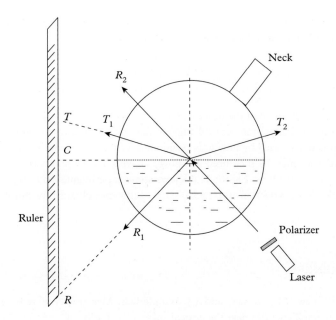

Figure 41.3 *Light through equatorial perimeter.*

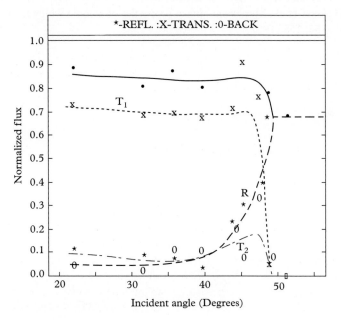

Figure 41.4 *Experimental data for direct reflection R, direct transmission T_1, and back reflection T_2 – see Fig. 41.3. The solid line is the combined flux of the three beams.*

Spherical flask setup

See Ref. [2]. A 1 l spherical flask is convenient, its neck positioned away from the vertical, so that light passes through the glass along the equatorial perimeter where glass inhomogeneities are smaller – see Fig. 41.3. If the center of the flask is O, then the refractive index is given by geometry and Snell's law as

$$n = \left(\frac{CO^2 + CR^2}{CO^2 + CT^2}\right)^{1/2} \frac{CT}{CR}. \tag{41.11}$$

A direct measurement of the distance OC is needed. One way to locate the point C is to point the laser at the critical angle so that the beam leaves horizontally. O's position is determined by judicious positioning of the laser beam vertically. The light intensity may be measured as above, or with a CdS photocell, calibrated by using suitable filters. Figure 41.4 shows some experimental data, reproduced from [2], for the reflection and transmission coefficients obtained by this method.

....

REFERENCES

1. D.E. Shaw, M.J. Hones, and F.J. Wunderlich, Measurents of reflected and transmitted energies near the critical angle, *Am. J. Phys.* **42**, 561 (1973).
2. J.H. Jones et al., Measuring optical properties of a liquid with a laser, *Am. J. Phys.* **50**, 158 (1982).

Polarization by transmission

42

Introduction	221
Theoretical ideas	221
Goals and possibilities	223
Experimental suggestions	224

Introduction

This project takes the student beyond the laws of Malus and Brewster, which are the usual topics in introductory optics laboratories on the subject of polarization. Of the commonly occurring ways of obtaining various degrees of polarizations (reflection, transmission, and scattering) from an unpolarized source, this project concentrates on transmission through a series of glass plates. The project assumes that the researcher will have the support of a mechanical workshop, to help build the apparatus to the specifications planned and designed by the researcher. There is opportunity here to gain a proper understanding of the Fresnel equations and the role of boundary conditions in the passage of light between two media.

Theoretical ideas

See Ref [1]. for full details of the theory.

Degree of polarization

Let unpolarized light be incident at an angle θ_0 to a glass plate, and $E_{\perp 0}, E_{\parallel 0}$ be the incident electric field vectors perpendicular to and parallel with the plane of incidence. The appropriate incident intensities are $I_{\perp 0}, I_{\parallel 0}$, while those transmitted are $I_\perp, I_\parallel$. The corresponding transmittances are $T_\perp = I_\perp/I_{\perp 0}$ and $T_\parallel = I_\parallel/I_{\parallel 0}$. Then the degree of polarization Δ may conveniently be defined by:

$$\Delta = (T_\perp - T_\parallel)/(T_\perp + T_\parallel). \tag{42.1}$$

Defined this way, $\Delta < 0$, since $T_\perp < T_\parallel$ for transmission.

Calculating Δ for incidence at the Brewster angle $\theta_0 = \theta_B$

Let an unpolarized beam of unit height and of width d_A in air be incident on a glass plate of refractive index n. In the plate, the beam width is d_P and the angle of refraction is θ_r, as shown in Fig. 42.1. From left to right, label the three regions air–glass–air by 0, 1, and 2. The incident unpolarized beam is represented by equal instantaneous perpendicular field components $E_{\perp 0} = E_{\parallel 0}$ and $I_{\perp 0} = I_{\parallel 0}$.

At the air–glass interface on the left, let $E_{\perp 0}$, $E'_{\perp_1}$, and $E''_{\perp_0}$ be the incident, refracted, and reflected fields, respectively. Consider an infinitesimally narrow rectangular path straddling the interface. Then by Faraday's law for the perpendicular component $E_\perp$, one has $\oint E \cdot ds = 0$. This means that

$$E_{\perp_0} = E'_{\perp_1} + E''_{\perp_0}. \tag{42.2}$$

By energy conservation we also have

$$E^2_{\perp_0} d_A = (E''_{\perp_0})^2 d_A + n(E'_{\perp_1})^2 d_P. \tag{42.3}$$

At the Brewster angle, $\theta_B + \theta_r = \pi/2$ and $\tan\theta_B = n$, so that Snell's law and the geometry of the figure lead to $d_P/d_A = n$. Substituting this into (42.3) and combining with (42.2) gives:

$$E'_{\perp_1} = 2E_{\perp_0}/(n^2 + 1). \tag{42.4}$$

At the glass–air interface on the right, a refracted beam $E'_{\perp_2}$ exits into air and $E''_{\perp_1}$ is reflected back into the plate. Then, instead of (42.2) and (42.3), we have:

$$E'_{\perp_1} + E''_{\perp_1} = E'_{\perp_2}, \tag{42.5}$$

$$(E'_{\perp_2})^2 d_A + n(E''_{\perp_1})^2 d_P = n(E'_{\perp_1})^2 d_P. \tag{42.6}$$

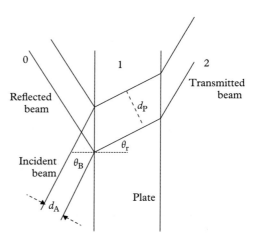

Figure 42.1 *Incidence at Brewster angle.*

From (42.5) and (42.6), and then using (42.4), we find:

$$E'_{\perp_2} = 2n^2 E'_{\perp_1}/(n^2 + 1) = 4n^2 E_{\perp_0}/(n^2 + 1)^2. \qquad (42.7)$$

Squaring (42.7), using $I_{\perp 0} = I_{\| 0}$, and with $T_\| = 1$ [since at θ_B there is no reflected ray with $E_\|$ at the first surface, one can argue that $(T_\|)_{\text{plate}} = 1$], we obtain, for a single plate,

$$T_\perp/T_\| = [2n/(n^2 + 1)]^4,$$

while for a stack of m parallel plates,

$$T_\perp/T_\| = [2n/(n^2 + 1)]^{4m}, \qquad (42.8)$$

a result which can also be proved from Fresnel's equations. Substituting (42.8) into (42.1) will then give Δ.

Arbitrary angle of incidence θ_0

For transmission through $2m$ surfaces one has [2]:

$$T_\perp/T_\| = \cos^{4m}(\theta_0 - \theta_r). \qquad (42.9)$$

This formula does not take into account light rays that are reflected inside the glass plates and therefore the Δ measured experimentally would be lower than the theory of (42.9) and (42.1) combined. Formulas correcting for the reflections may be found in [2]. The correction factor multiplying (42.9) is given by

$$(1 - R^2_\|)/(1 - R^2_\perp), \qquad (42.10)$$

where $R_\perp = [\sin(\theta_0 - \theta_r)/\sin(\theta_0 + \theta_r)]^2$, $R_\| = [\tan(\theta_0 - \theta_r)/\tan(\theta_0 + \theta_r)]^2$.

Goals and possibilities

- Measure the refractive index n of the glass plates used.
- Using Polaroid sheets, measure the slight degree of polarization of the light source to be used in the experiments, by calculating the ratio $I_\perp/I_\|$, where $I_\perp$ and $I_\|$ are measured by a photodiode. This is needed in order to calculate properly the values of Δ.
- Construct an apparatus that can measure the degree of polarization of the light exiting a variable number of small glass plates.

- Check that the photodiode current is directly proportional to the light intensity. This can be done by a Malus's law measurement. To take into account the polarization dependence of the photodetector response, either experiment with it yourself, or refer to its technical data.
- Determine the Brewster angle θ_B experimentally (disappearance of reflected ray) and also from $\theta_B = \tan^{-1} n$.
- Measure the degree of polarization Δ for a series of parallel plates when the angle of incidence is θ_B; plot Δ versus the number of polarizing plates. Compare with (42.8) and (42.1).
- Measure the degree of polarization Δ for a series of parallel plates for various angles of incidence; plot Δ versus the number of polarizing plates. Compare with (42.9), (42.10), and (42.1).
- Repeat the measurements using plates with different n (different glass or plastic).

Experimental suggestions

See Ref. [1]. An adequate light source is a small 12 V 50 W halogen car lamp. Analyzers can be good quality Polaroid squares. Since these do not affect IR radiation, mount an IR filter in front of the source. In addition, the photodiode which is used to detect the intensity of light emerging from the plates should have a 700 nm cut-off. As the photodiode is some 5 mm square, use collimator slits to define the beam from the source. The glass plates to be used as polarizers by transmission can be microscope slides, very carefully cleaned and wiped with lens wipers in order to prevent scattering by dust particles. To avoid the need for amplification the source–detector distance should not exceed a few tens of centimeters.

It is very important to align accurately all the components in a collinear manner on optical benches on either side of the turntable. The latter could be taken from a spectrometer with the arms removed, allowing rotation angles to be measured accurately. Other arrangements should allow angle measurements to an accuracy of at least 0.5°. Mounting the Polaroid sheets on graduated rotating mounts, found in optical bench kits, will ensure accurate analyzer alignment, with the polarizing axes of the two sheets perpendicular to each other.

The main technical problem to be solved is the mounting of a number of glass plates accurately next to and parallel with one another, at the center of the turntable. For this purpose a mount with parallel grooves should be machined at a mechanical workshop, with a mark at its center, and a pointer to the normal to its plane, to facilitate accurate alignment with the other components.

Alignment errors may be compensated by taking detector readings when the stack of plates is rotated so as to give incidence of the incoming beam both at

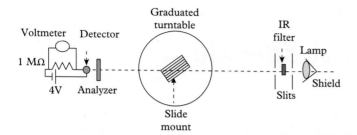

Figure 42.2 *Experimental setup.*

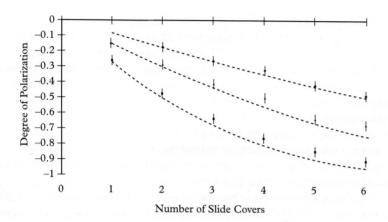

Figure 42.3 *Degree of polarization as a function of the number of slides.*

$+\theta_0$ and $-\theta_0$, and then averaging. The small lateral shift of the transmitted beam means that for a strong signal one should displace the analyzer laterally by a few millimeters. The setup is shown schematically in Fig. 42.2.

The graph in Fig. 42.3, after Ref. [1], shows the degree of polarization as a function of the number of glass plates from 1 to 6, for various values of the angle of incidence: 45°, 56°, 70°.

....

REFERENCES

1. R.E. Benenson, Light polarization by transmission, *Eur. J. Phys.* **21**, 571 (2000).
2. J.M. Bennett and H.E. Bennett, Polarization, in *Handbook of Optics* I, eds W.G. Driscoll and W. Vaughan (McGraw-Hill, New York, 1978), pp. 88–96.

43 Laser speckle

Introduction 226
Theoretical ideas 226
Experimental suggestions 227

Introduction

A random pattern within the intensity of coherent light can be seen in many situations, for example when a laser is scattered off a rough surface, as shown in Figs 43.1 and 43.2. This random "noise" pattern of interference is termed "speckle."

Speckle also occurs in a highly magnified image of a star through the atmosphere or through imperfect optics; when sunlight is scattered by a fingernail; when radio waves are scattered from rough surfaces such as ground or sea; and in ultrasonic imaging. When the speckle pattern changes in time due, for example, to changes in the illuminated surface, we have dynamic speckle. This can be used to monitor motion and activity, for example in an optical flow sensor, blood flow dynamics, or in holographic interferometry. Further uses are in eye testing, stellar speckle astronomy, and image processing.

The aim of this project is to become familiar with the basic phenomenon, and to understand the background to the widespread use of laser speckle. The project proceeds in several stages:

1. Qualitative, subjective observations. (The terms *subjective* and *objective* are explained below.)
2. Simple quantitative measurements of speckle properties.
3. Laser speckle photography and the study of motion by high resolution digital cameras.
4. An application to the motion caused by camera shake during exposure.

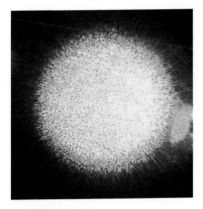

Figure 43.1 *Subjective image of green laser pointer speckle.*

Figure 43.2 *Objective speckle pattern for laser scattered from plastic surface.*

Theoretical ideas

Laser speckle will result from interference of coherent light waves scattered from a diffuse surface whose roughness is of the order of, or larger than, the wavelength. It can be observed both in the Fraunhofer (far-field) mode, using a lens to create an image, or the Fresnel (near-field) mode, existing as a three-dimensional pattern in front of the scattering surface. When we observe an illuminated surface, we detect the average energy of the light at the surface; thus the brightness of a given point on a surface, which has been illuminated by a set of random scatterers with a single frequency, is constant over time, but varies randomly from point to point; i.e., it is a speckle pattern. If light of low coherence (i.e. made up of many wavelengths) is used, a speckle pattern will not normally be observed, because the speckle patterns produced by individual wavelengths have different dimensions and will normally average one another out. However, speckle patterns can be observed in polychromatic light under some conditions.

The size of the speckles is a function of the wavelength of the light, the size of the laser beam that illuminates the first surface, and the distance between this surface and the surface where the speckle pattern is formed. This is because intensities become uncorrelated when the relative path difference between light scattered from the center of the illuminated area compared with light scattered from the edge of the area changes by λ. A useful expression for the mean speckle size is $\lambda z/L$, where L is the width of the illuminated area and z is the distance between the object and the location of the speckle pattern.

The theory and many applications are described in [1]. The book by Francon [2] is a useful introduction to the subject.

Experimental suggestions

Qualitative subjective observation

(a) Mount a 1 mW He–Ne laser on an optical bench, expand its beam by a microscope objective and view the speckle pattern with one eye as the beam falls on screens of various surface textures (white card, various grades of sandpaper, matt aluminum). No major differences are discerned between the surfaces, as long as the surface roughness exceeds the wavelength. Move the screen from side to side: this motion destroys the speckle. Move your head from side to side: this causes the pattern to shift. If wearing spectacles, remove them: the speckle pattern is not blurred out by defocussing. Observe eye safety measures at all times.

(b) Introduce a polarizer into the beam and observe through a second polarizer: the white card and the aluminum will appear different. Light penetrates into the card surface, causing multiple scattering which destroys the polarization. Single scattering on the aluminum surface does not.

(c) View through a series of pinholes punched into a black card: the smaller the hole, the coarser the pattern. View through a variable-width slit: speckles spread out into lateral streaks whose lengths vary in inverse proportion to the slit width. If the speckle pattern is photographed, the image graininess will depend on the lens f-number – diffraction limits the detail seen to a size larger than the speckle. View through a pinhole, but replace the screen by a human hand: the speckle pattern is disturbed by the continuous movement of the skin.

(d) Replace the screen by a glass or plastic cell containing a suspension of talcum powder in water, and view the speckle pattern by looking at the light transmitted through the cell, using a pinhole held close to the eye so as to make the speckles look larger: the observed speckle is seen to "boil," caused by multiple scattering from the moving talc particles – Brownian motion of micron-sized particles. Indeed, laser speckle has been used to study Brownian motion in many contexts – one example is Ref. [3].

The change in speckle size with aperture and the positioning of the imaging system may be explained qualitatively as follows. Each point in the image can be considered to be illuminated by a finite area in the object. The size of this area is determined by the diffraction-limited resolution of the aperture, which is given by the Airy disk whose diameter is $2.4\lambda u/D$, where λ is the wavelength of the light, u is the distance between the object and the lens, and D is the diameter of the aperture. The light at neighboring points in the image has been scattered from areas that have many points in common and the intensity of two such points will not differ greatly. But notice that two points in the image that are illuminated by areas in the object which are separated by the diameter of the Airy disk will have light intensities that are unrelated. This corresponds to a distance in the image of $2.4\lambda v/D$, where v is the distance between the aperture and the image. Thus, the size of the speckles in the image is of this order. The change in speckle size with aperture can be observed by looking at a laser spot on a wall directly, and then through a very small hole. The speckles will be seen to increase significantly in size.

Objective measurements

(a) Direct the laser beam, unexpanded, onto a ground-glass diffuser sheet. This forward-scatters the light onto the screen. A coarse-grained speckle is seen. Move the screen towards and away from the ground glass: the pattern scale will change, as shown by measuring the distance between two adjacent speckles on the screen as a function of different screen–ground glass distances. The relation is linear. If a second ground-glass screen is brought up close to the first one, the projected pattern will have reduced speckle size, due to the second scatterer being illuminated over a larger area.

(b) Focus the laser beam onto the ground glass with various microscope objective lenses mounted on a rack and pinion holder. As focus is approached, the speckle pattern grows coarse. At focus, the highest power objective (say ×20) gives a much larger speckle than the lowest one (say ×3), because it focuses to a smaller beam diameter. The spot diameter D can be expressed in terms of the speckle size s and the screen–diffuser distance L: $D = 1.2\lambda L/s$, where the speckle size is taken to be the distance between two adjacent bright speckles.

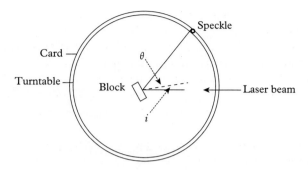

Figure 43.3 *To observe speckle rotation.*

(c) When the rough scattering surface is rotated, the speckle pattern rotates as a whole. This may be observed by the following method: focus the laser with a ×3 microscope objective on to an aluminum block which is mounted on a calibrated turntable whose axis of rotation lies along the surface of the block, as shown Fig. 43.3. The speckle pattern is projected onto a card, graduated in degrees, which is fixed all around the perimeter of the turntable. With the laser focused on a point lying on the axis of rotation, a single speckle can be followed as the table is rotated. Choose a speckle close to the direction of the incident laser beam, and record the angle of speckle rotation ϕ as a function of table rotation angle i. A linear relation will be found, namely, $\delta\phi = 2\delta i$, as if the aluminum surface were acting as a plane mirror. However, if one chooses a speckle at an angle θ to the beam direction, the relation will not be linear, and the rough surface behaves as though it were a reflection grating: $\delta\phi = [1 + (\cos i/\cos\theta)]\delta i$. It's as though the speckle pattern is being generated by a collection of diffraction gratings of varying line spacings and orientations.

If now the aluminum block is displaced, so that its surface does not include the rotation axis, the speckles will not only appear to rotate with the table, but change their brightness and structure. This is due to the gradual change in the area being illuminated.

Speckle photography

Figure 43.4 shows the experimental arrangement, based on the pioneering work of Burch and Tokarski [4]. The laser beam is expanded using a microscope objective/pinhole combination to illuminate a fine-ground-glass diffuser screen. A circular aperture of variable diameter D placed near the diffuser enables one to vary the speckle size, which is of the order $\lambda H/D$, where H is the distance between the diffuser and the image plane (photographic emulsion in the original film cameras). It is suggested that $H \approx 800$ mm and $D < 60$ mm, so that for $\lambda \approx 633$ nm

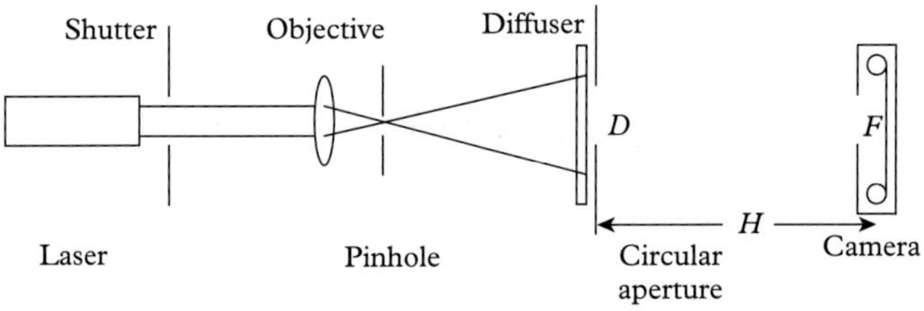

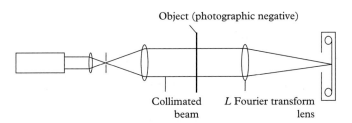

Figure 43.4 *Setup for speckle photography.*

Figure 43.5 *Recording diffraction patterns or reconstruction of image motion.*

the speckle size is > 10μm. Either an external shutter, a cable release, or a self-timer on a digital SLR camera should be used, together with an optical bench and a sturdy table, in order to reduce vibration effects. The diffuser screen should be mounted on a micrometer-controlled platform which enables one to shift it in its own plane by steps of 0.1 mm or smaller.

With the setup of Fig. 43.4, record the speckle pattern by choosing a suitable exposure time (depends on your camera settings).

Then shift the diffusing screen in its own plane by d (in the range 0.1–0.5 mm), and make a second exposure (same time exposure) on the same frame. Two identical speckle patterns shifted by d are recorded. Print the double exposure on a slide made from a transparency. This slide, when it is illuminated in the setup of Fig. 43.5, will give rise to a fringe pattern in the back focal plane of lens L; a convenient focal length is 90 mm (in the theory of Fourier optics this is called the Fourier transform lens; for a tutorial on this topic see [5]). The intensity distribution is proportional to $\cos^2(\pi dx/f_1\lambda)$, where f_1 is the focal length of the lens L. The two speckles act as identical coherent sources, and so produce Young-type fringes in the diffraction plane. The fringe separation c in the x direction will be given by $c = (\lambda/d)$ – see [2]. If $d = 0.5$ mm, this separation amounts to 0.1 mm.

It is now worth exploring systematically the following variation of parameters:

1. The effect of increasing d, resulting in a reduction of fringe spacing.
2. The effect of increasing the number of exposures from two, by shifting the diffuser in several equal steps d. This results in a sharpening of the fringes, and in an increase in the number of intermediate maxima.
3. The effect of unequal exposure times, while preserving equal displacements. A rearrangement of the intermediate fringes is expected.

4. The effect of unequal displacements, resulting in a rich fringe texture modulated by new periodicities.

It is interesting now to reconstruct the original speckle displacement, using the same setup as in Fig. 43.5, as follows: Prepare a slide from the interference fringes – this is equivalent to a grating of spacing c. Illuminate this slide by a plane monochromatic wave at normal incidence, as in Fig. 43.3. The diffraction pattern is then recorded in the focal plane of a lens of long focal length f_2 (500 mm), chosen so as to give a magnification f_2/f_1 of the speckle displacement. The result is a pattern of diffraction spots in various orders, whose separation is of the order df_2/f_1.

Camera shake

The double exposure method to show movement, as quantified by the reconstruction method, can be applied to show the effect of shake by a hand-held camera. For this, use a digital camera that does *not* have a stabilizer built into it. One needs to take two photographs of the speckle pattern, once with the camera firmly clamped to its mount, and once, the diffuser screen having been shifted by some 0.5 mm, with the camera firmly held by hand on its mount. As an aid to the alignment for the two exposures, a horizontal and two vertical sighting lines (0.5 mm apart) may be marked on the diffuser. These are then observed through the camera lens, at a magnification close to unity. Sighting errors will result in fringes inclined to the vertical, whereas motion during exposure due to hand shake will result in broad bands shifted with respect to each other. The movements, whether in the vertical or horizontal direction, will be seen clearly by means of double spots in the reconstructed diffraction pattern.

....

REFERENCES

1. J.C. Dainty (ed.), *Laser Speckle and Related Phenomena* (Springer, Berlin, 1984).
2. M. Francon, *Laser Speckle and Applications in Optics* (Academic, New York, 1979).
3. M. Qian, Q. Yan, X-W. Ni, and H-R. Zheng, Detection of nanoparticle Brownian motions in a nanofluid using laser speckle velocimetry, *Lasers in Engineering* **20**, 117 (2010).
4. J.M. Burch and M.J. Tokarski, Production of multiple beam fringes from photographic scatterers, *Optica Acta* **15**, 101 (1968).
5. University of Warwick, The lens as a Fourier transform system, online at <http://www.eng.warwick.ac.uk/~espbc/courses/undergrad/lec8/fourier_lens.htm>.

44 Light scattering from suspensions

Introduction 232
Theoretical ideas 232
Goals and possibilities 233
Experimental suggestions 234

Introduction

Phenomena like critical opalescence, fog and mist, and the color of the sky and lakes at different times of the day can all be explained in terms of light scattering. Light scattering in its many forms is extremely important as a research and diagnostic tool in physics, chemistry, and biology, as well as in applied research and industrial applications [1–5]. To name but a few applications, light scattering experiments have been carried out in red blood cells, paints, suspensions in wines and soft drinks, microspheres, and in the atmosphere. The theoretical treatments (associated with the names of Rayleigh, Mie, Gans, Raman, and Brillouin) depend on λ/a, the ratio of the wavelength and the radius of the scattering particles.

The project proposed here provides an introduction to the subject by measuring the scattered light intensity as a function of angle (differential light scattering, DLS for short) for aqueous solutions of polystyrene microsphere suspensions of known standard sizes. The experience gained thereby is then used to study the nature of suspended scattering particles in various brands of beer [6], each having a characteristic DLS signature.

Theoretical ideas

Consider randomly dispersed spheres of refractive index n_s and radius $a << \lambda_s = \lambda_0/n_s$, immersed in a medium of refractive index n_m, scattering incident light of wavelength λ_0, shown schematically in Fig. 44.1. Under these conditions one has Rayleigh scattering with the characteristic $1/\lambda^4$ factor: Every point within a single sphere is assumed to be equidistant from the point of observation and to be driven in phase by the incident light, so that the scattered amplitude is proportional to the sphere's volume V, and the intensity to V^2. For N incoherent scatterers per unit volume, the total scattered intensity is proportional to N. If the sample is small, and the distance r from the scatterer to the observation point is large, we have the following expression [5] for the scattered intensity per unit volume as a function of scattering angle θ:

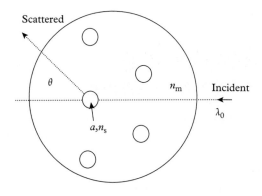

Figure 44.1 *Randomly dispersed spheres of radius a and refractive index n_s in a fluid of refractive index n_m.*

$$I_R = \frac{(9\pi^2/2)I_0 N V^2}{r^2 \lambda_m^4}(1+\cos^2\theta)[(n_r^2-1)/(n_r^2+2)]^2, \qquad (44.1)$$

where the incident intensity is I_0, $n_r = n_s/n_m$, and $\lambda_m = \lambda_0/n_m$.

For larger spheres, different phases of the wave from different parts of the sphere must be taken into account. A simple treatment is due to Rayleigh and Gans [3–5], and is more accessible than the full theory of Mie. It is valid when $(n_r - 1) < 1$, and when the phase shift, namely $4\pi a(n_r - 1)/\lambda_m < 1$, is small. In this approach one uses the square root of expression (44.1) for the amplitude from each elementary volume dV, multiplied by an appropriate phase factor $\exp(i\delta)$, then integrates over the volume, and finally squares for the final intensity. The phase factor resulting from the optical path differences for different parts of the sphere is obtained by dividing the sphere up into "slices" which form planes at some angle to the incident beam, each slice producing the same phase in the scattered beam (somewhat like the Bragg planes in X-ray diffraction). This is then followed by integration over the sphere. The resulting scattered intensity is [5]:

$$I_{RG} = \frac{(4\pi^2)I_0 N V^2}{r^2 \lambda_m^4}(1+\cos^2\theta)(n_r^2-1)^2[(\sin u - u\cos u)/u^3]^2, \qquad (44.2)$$

where $u = (4\pi a/\lambda_m)\sin(\theta/2)$.

Goals and possibilities

- Experiment with aqueous solutions of polystyrene microspheres of such dilution that single-particle scattering is relevant: determine the range over which the scattered intensity varies linearly with concentration.
- Use several particle sizes, having radii between 0.05 and 0.25 μm.

- Measure the scattered intensity as a function of angle for different-sized particles. Try to fit the data with the theoretical expressions given in (44.1) and (44.2).
- Different concentrations of standard spheres could be combined to simulate polydispersive systems.
- Carry out DLS experiments on several brands of beer, or fruit juice. Can the data be fitted by either of the above theories using a single size for suspension particles, or is a combination of different sizes necessary?

Experimental suggestions

Light from a He–Ne laser of wavelength 0.63 μm is used. To gain experience in interpreting DLS measurements, it is useful to begin the project by scattering off aqueous solutions of polystyrene microspheres whose sizes and refractive indices are known. Ordering a range of sphere diameters from the suppliers, say 0.05, 0.13, 0.25 μm, will enable one to cover and test both the Rayleigh and Rayleigh–Gans theories – both (44.1) and (44.2). Hold the solution samples in a small cylindrical glass cuvette placed on the platform of a spectrometer table, as shown in Fig. 44.2. Laser light scattered by the sample at any angle θ is collected

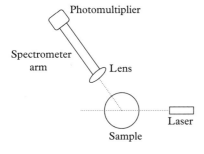

Figure 44.2 *Experimental setup.*

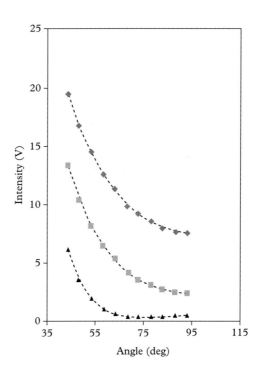

Figure 44.3 Diamonds: $a = 0.055$ μm, $N = 2\times10^{10}$ ml^{-1}; squares: $a = 0.128$ μm, $N = 1\times10^{8}$ ml^{-1}; triangles: $a = 0.25$ μm, $N = 2.5\times10^{8}$ ml^{-1}. Fitted with Rayleigh–Debye theory, dotted curves. Reproduced from [6].

by a lens and focused on a photomultiplier mounted on the rotating arm of the spectrometer. Its output to any recording device is proportional to the scattered light intensity. When fitting the experimental results to the theoretical formulae, account must be taken of the variation of solid angle with θ into which the light is scattered. To fit data points to the theoretical curves you will need to normalize them by a common factor.

Figure 44.3 shows the scattered intensity as a function of angle for microspheres of three different sizes. Suspensions in beer [7] are neither spherical nor are they of a single size, being in the range 0.2–0.4 μm. It is nevertheless of interest to carry out DLS measurements on various brands, in the manner described above. One might be able to fit the results by combining the above expressions for the intensity with two particle sizes in the expected range. Perhaps each brand will have its own DLS "signature." It may be of interest to see how this signature varies with time (particle sizes vary with time).

REFERENCES

1. B.J. Berne and R. Pecora, *Dynamic Light Scattering with Application to Chemistry, Biology and Physics* (Wiley, New York, 1976).
2. D.A. Long, Survey of Light-scattering Phenomena, in *The Raman Effect: A Unified Treatment of the Theory of Raman Scattering by Molecules* (Wiley, New York, 2002), ch. 1.
3. H.C. van de Hulst, *Light Scattering by Small Particles* (Wiley, New York, 1975).
4. C.F. Bohren and D.R. Huffman, *Absorption and Scattering by Small Particles* (Wiley-VCH, Weinheim, 2008).
5. M. Kerker, *The Scattering of Light and Other Electromagnetic Radiation* (Academic, New York, 1969).
6. M.E. Bacon, W.J. Johnson, and M.A. Day, Experiments in differential light scattering of fluids, *Eur. J. Phys.* 7, 259 (1986).
7. R. Thorne and K. Swendsen, Particle size of beer turbidigens and its influence on nephelometry, *J. Inst. Brew.* 68, 257 (1962).

45 Light intensity from a line source

Introduction 236
Theoretical ideas 236
Goals and possibilities 237
Experimental suggestions 237

Introduction

This project investigates the dependence of the light intensity from a linear light source on the distance along the axis perpendicular to the source. The theoretical background to the experiments exploits the analogy between charge distributions and light-source distributions. Of interest is the transition from near-field to far-field behavior, the efficiency of the source as a function of the power supplied to it, and the effect of the angular dependence of the radiation detector response on intensity measurements.

Theoretical ideas

See Ref. [1]. The electric field on the perpendicular bisector of a uniformly distributed line charge of length L, at distances ranging through very near, near, and far, is a constant for distances $x \ll L$, is $\propto 1/x$ for $x < L$, and behaves as $1/x^2$ for $x \gg L$, respectively.

The power per unit area from an incoherent unpolarized light source, the irradiance, is an integral over the length of the source from elementary point source contributions. At a distance r from a point source the irradiance is given by the time-averaged Poynting vector [1]

$$\overline{S} = P/4\pi r^2, \qquad (45.1)$$

where P is the radiated power. Equation (45.1) is just the electric field of a charge of size $\varepsilon_0 P$. Therefore one may take over the mathematical theory from electrostatics in order to calculate the integrated intensity distribution due to a linear light source whose parameters are shown in Fig. 45.1. So the irradiance at a point D on the axis will be given by:

$$\overline{S}(x) = \frac{P}{2\pi x (L^2 + 4x^2)^{1/2}}. \qquad (45.2)$$

Let a detector of surface area A be placed at D, and assume that $\sqrt{A} \ll x$, so that the Poynting vector may be considered a constant across A. Let θ be the

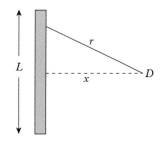

Figure 45.1 *A linear light source.*

angle between the normal to A and the axis. Then the power detected on the perpendicular bisector will be given by

$$\overline{S}_{\det}(x) = \frac{PA|\cos\theta|}{2\pi x(L^2 + 4x^2)^{1/2}}. \tag{45.3}$$

In the limit $x \gg L$ we have a $1/x^2$ dependence, while for $x < L$ the irradiance behaves as $1/x$. The $\cos\theta$ dependence is not usually borne out by experiment, due to the limitations and characteristics of the various quantum detectors in use. Most of these cannot "see" beyond a maximum angle, and this fact is especially important in the near-field limit of (45.3), when parts of the line source will make large angles with the detector surface. For this reason, (45.3) should be combined with the actual angular response of the detector, determined experimentally.

Goals and possibilities

- Measure the angular response of the particular detector used to detect power.
- Measure the x dependence of the power for the full range of distances, from the very small to the very large, in order to be able to compare with the theoretical predictions of the transition from near-field to far-field dependence.
- Use single-filament line sources of various lengths, 20–50 cm, in order to check how the transition from near- to far-field behavior is affected by size.
- Use a thermopile to explore the wavelength dependence of the (gray-body) radiation from the source.

Experimental suggestions

See Ref. [1]. Very careful vertical mounting of the filament lamp should make sure that the filament is perpendicular to the horizontal axis along which the detector will be located and moved.

The photodetector (photodiode, phototransistor, etc.) should be mounted on an optical bench. Mounting it on a component holder with a swivel joint enables you to measure the angle dependence of the detector signal easily. The output of the sensor, suitably amplified, can be fed to any data acquisition system.

If the lamp has a transparent glass envelope, careful attention must be paid to internal reflections and stray light. These can be eliminated by a mask slit placed in front of the lamp. There are thin linear incandescent lamps some 50 cm long (used for bathroom lighting) which look like fluorescent lamps. These could serve as excellent line sources. Either the Osram-Sylvania "Linestra" 60 W, or the

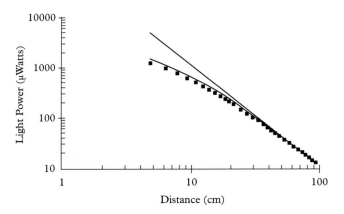

Figure 45.2 *Log-log plot of light intensity as a function of distance, showing the transition from near- to far-field behavior. The smooth curve is the modified cosine detector response; the top line is the inverse square law for comparison.*

Philips "Philinea" models are suitable. Shorter bulbs with glass envelopes, some 25 cm long, can be obtained from various supply houses.

Using a variable voltage source one could vary the intensity of the lamp and study its efficiency as a function of the power input.

If one wishes to extend the project both in time and scope, one could combine this project with that for the lamp burnout described in Chapter 9.

Figure 45.2, reproduced from [1], shows the near to far field transition on a logarithmic scale. It has also a $1/r^2$ line for comparison.

REFERENCES

1. X. Yan, Y. Yu, L. Shen, and K.H. Wanser, Near-field to far-field transition of a finite line source, *Am. J. Phys.* **63**, 47 (1995).

Light interference in reflecting tubes

46

Introduction	239
Theoretical ideas	239
Goals and possibilities	240
Experimental suggestions	241

Introduction

Multiple reflections of light incident on the reflecting interior of a tube give rise to well-known symmetrical light rings, a phenomenon used to check the accurate alignment of gun barrels. Similar phenomena occur for light reflected between two parallel mirrors. The various interference phenomena are similar to Lloyd's mirror. This project is devoted to detecting and quantitatively recording the interference patterns that arise, and explaining their origin.

Theoretical ideas

See Ref. [1]. Figure 46.1 shows parallel mirrors separated by D, or a cross section of a cylindrical tube of inner diameter D and length L. The point source O is at a distance K from one end of the reflecting tube. The screen S is at a distance mL from the end of the tube or distance $K + L + mL$ from the source. Shown are rays from the source reflected on the inner walls at the beginning and end of the tube. After relection they appear to originate from virtual sources O' and O''. Simple geometry and trigonometry, using the various triangles formed, shows the following results.

1. The beams of light reaching the screen S directly, without reflection from the walls of the tube, form a spot of diameter $D[1 + mL/(K + L)]$. For parallel mirrors a rectangular strip of this width is formed.

2. A single reflection at the near end of the tube has rays exiting at the far end, appearing to originate from O' and O''. These rays intersect at a distance $2(K + 1)/3$ from O, at O_2, which is the apex of a cone of outer base diameter $MT = D[1 + 3mL/(K + L)]$. The inner diameter NR of this ring of light on S subtends an angle $\theta = BOO_2$ given by $\theta = \arctan[D/2(K + L)]$.

3. The rays reflected at B and B' intersect at O_1, where $OO_1 = 2(K + L)$, and enclose an area on S of diameter $PQ = D(1 - mL)/(K + L)$.

4. Consider now the various possible placements of the screen S.

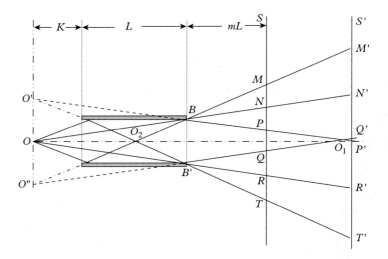

Figure 46.1 *Geometry of reflections between parallel mirrors or a cross section of a cylindrical tube. Source at O, tube length L, inner diameter D. Singly reflected rays produce virtual sources at O′ and O″. Screens S (and S′) at distance mL are shorter (longer) than 2(K + L) and show the possibility of two- or three-beam interference regions.*

(a) If screen S were to be placed at O_1, the inner diameter of the area illuminated by singly reflected rays would just reduce to a point O_1.

(b) If S were placed to the left of O_1 one would discern the following three areas on it:

   (i) A ring, MN and RT, due to single reflections.

   (ii) A ring NP and QR. This is a region where direct and reflected beams can interfere with each other.

   (iii) A circular disk where beams from the source O and those from the image ring of the source in the tube can superpose and interfere.

(c) If S is to the right of O_1, the third area in (b) is only illuminated by direct rays from O.

5. By geometry, the number of possible reflections in the tube is the integer part of $(1 + L/2K)$.

So much for geometrical optics. But there will also be diffraction at the edges B, B′. This will produce diffraction patterns on the screen.

Goals and possibilities

- Explore interference both from two parallel mirrors and from a tube with a reflecting surface.
- Study the patterns formed when the screen is placed at different positions.

- Separate the beams coming from the source and those coming from the image circle of the source in the tube.
- Photograph or record by other means the diffraction pattern from the end of the tube, and, separately, the illuminated area created by reflection from the walls of the tube. Detect the intensity of light in all cases.
- Use various diaphragms, stops, and lenses to investigate the variation in size, separation, and intensity of the patterns.

Experimental suggestions

Use as the point source a spatially filtered He–Ne laser. The pinhole of the filter is about 10 μm in diameter. The first Airy disk, up to the first minimum of the diffraction pattern of the pinhole, will serve as the point source. A tube of length 4 cm and diameter 0.3 cm is convenient. For the parallel mirror experiment, use aluminum-coated mirrors 3 cm long and spaced 1 mm apart. The experimental setup is shown in Fig. 46.2.

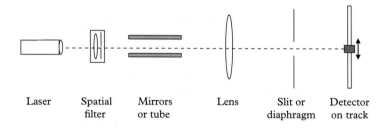

Figure 46.2 *Experimental setup. Spatial filter acts as a point source, diaphragm is beam selector.*

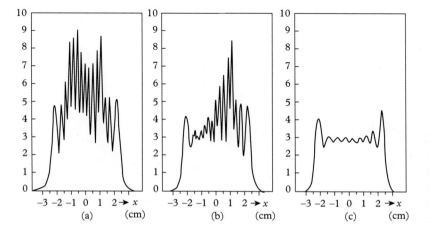

Figure 46.3 *Intensity distributions: (a) wide-open slit, all three beams reach screen; (b) superposition of direct and one reflected beam, the second reflection being blocked by the slit; (c) only direct beam, diffraction from edges B and B'.*

The interference and diffraction pattern may be recorded with a photodiode scanning the screen plane in controlled steps, or a digital CCD camera.

A converging lens of 15 cm focal length is useful in two ways. It increases the fringe separation; in combination with a diaphragm positioned in front of the plane image of the source, it also separates the diffraction pattern formed by the tube edge BB' from the illuminated area whose source is the reflection from the wall of the tube.

Figure 46.3, reproduced from [2], shows the distribution of intensity in the plane of the photodiode with lens and diaphragm.

....

REFERENCES

1. D.K Cohen and J.E. Potts, Light transmission through reflecting cylindrical tubes, *Am. J. Phys.* **46**, 727 (1978).
2. Y. Lion and Y. Renotte, Interference of light by reflection on the inner walls of cylindrical tubes, *Eur. J. Phys.* **13**, 47 (1992).

Part 6

Temperature and Heat

47	Cooling I	245
48	Cooling II	248
49	The Leidenfrost effect I	254
50	The Leidenfrost effect II: drop oscillations	258
51	The drinking bird	260
52	Liquid–vapor equilibrium	266
53	Solar radiation flux	270

Cooling I

Introduction

Newton's law of cooling is often cited when describing the rate of cooling of a body from its initial temperature T_m to the ambient temperature T_a. According to this law the time dependence of the temperature, $T(t)$, is an exponentially decreasing function. The aim of this project is to investigate the temperature range where the law describes experiments faithfully, and to suggest necessary corrections due to radiation losses at high temperatures.

Introduction	245
Theoretical ideas	245
Goals and possibilities	247
Experimental suggestions	247

Theoretical ideas

For the theoretical background, see Ref. [1].

Only radiation losses

We assume that the cooling body starts out red hot, so that radiation heat loss is dominant and other modes like conduction, convection, and evaporation are negligible. The rate of heat loss may then be written as

$$(dQ/dt)_{\text{rad}} = -\varepsilon \sigma A \left(T^4 - T_s^4\right), \qquad (47.1)$$

where ε is the emissivity, σ the Stefan–Boltzmann constant, A the effective radiating area, and T_s is some *effective* ambient temperature (sometimes used as a free fitting parameter, this represents the temperature "seen" by the effective radiating area). The first term represents the power radiated to the surroundings, while the second describes power absorbed from the surroundings. A loss of heat dQ from the body leads to a change dT in its temperature:

$$dQ = mc\,dT, \qquad (47.2)$$

c being the specific heat and m the mass of the body. Written in dimensionless form, (47.1) becomes

$$d\theta/d\tau = 1 - \theta^4, \qquad (47.3)$$

where $\theta = T/T_s$, $\tau = t/t_r$, and $t_r = mc/\varepsilon\sigma A T_s^3$ is a characteristic time constant for radiative cooling (hereafter called the "radiation time," it is analogous to the RC time constant familiar from discharging a capacitor).

The solution of (47.3) is certainly not an exponential function. It is separable, and its implicit solution is

$$\frac{\theta - 1}{\theta + 1} = \frac{\theta_m - 1}{\theta_m + 1} \exp\{-2(\tan^{-1}\theta_m - \tan^{-1}\theta)\} \exp(-4\tau), \qquad (47.4)$$

where $\theta_m = T_m/T_s$ and T_m is the initial temperature of the body.

One may consider the solution (47.4) in two different limiting cases.

T is near T_s

Then, from (47.3), we get:

$$\frac{d\theta}{d\tau} = (1 - \theta^4) = (1 - \theta)(1 + \theta)(1 + \theta^2) \approx -4(\theta - 1). \qquad (47.5)$$

The solution of (47.5) is a simple exponential function, corresponding to Newton's law of cooling.

T is very near T_m

In this case one obtains, after some algebra,

$$\theta = 1 + \frac{2(\theta_m - 1)(\theta_m^2 - 2\theta_m + 3)}{(\theta_m^2 + 1)[(\theta_m + 1)e^{4\tau} - \theta_m] - 4(\theta_m - 1)}. \qquad (47.6)$$

While more complex than the first limit, the exponential in the denominator still provides the rapid decay expected.

Losses due to both radiation and conduction

We now add conduction to (47.1). Suppose that the body loses heat through some "wall" of thickness d and area A. Heat flow due to conduction alone is described by the equation

$$\left(\frac{dQ}{dt}\right)_{\text{cond}} = -\frac{KA}{d}(T - T_m), \qquad (47.7)$$

where K is the thermal conductivity. We again transform this equation into dimensionless quantities, resulting in

$$d\theta/d\tau = 1 - \theta, \qquad (47.8)$$

in which $\tau = t/t_c$, $t_c = mcd/KA$, and t_c is a typical "conduction time" (this, too, is analogous to an RC time constant, here expressible in terms of the thermal resistance and the heat capacity). Equation (47.8) by itself is an expression of Newton's law, with solution

$$\theta = 1 + (\theta_m - 1)e^{-\tau}. \qquad (47.9)$$

Let us now join (47.3) and (47.8), to depict conduction and radiation together:

$$\frac{d\theta}{d\tau} = (1-\theta) + \gamma\left(1-\theta^4\right) = (1+\gamma) - \theta - \gamma\theta^4, \qquad (47.10)$$

where we defined $\gamma = t_c/t_r = \varepsilon\sigma d T_a^3/K$ as the ratio of conduction to radiation times. This equation may only be solved numerically, whereby γ will be an adjustable parameter chosen to fit the experimental data.

Goals and possibilities

- Heat a body until it is red hot. The body could be a heater or an electric cooker. The filament of an incandescent lamp is also suitable. Measure its temperature as a function of time.
- Try to fit the data to (47.5).
- Create conditions where both conduction and radiation are important (for some situations, this may amount to considering a broad range of temperatures, with conduction dominant in one limit and radiation in another), and measure $T(t)$. Try to fit the data to the numerical solution of (47.10).

Experimental suggestions

For the lamp, filament temperature is measured in terms of the resistance, as in Chapter 9.

For a heater or a stove one needs a digital thermocouple working in a suitable range.

A convenient system for observing both conductive and radiative losses is a piece of styrofoam into which you insert a thermocouple.

In the first part of the experiment wrap the styrofoam with shiny aluminum foil, so it does not radiate (a poor absorber is also a poor emitter). Immerse it in water, bring the water to the boil, remove the styrofoam and measure its temperature as a function of time, $T(t)$. In the second part, paint the foil matt black and repeat – now both conduction and radiation will be relevant. In this case the data should be compared with the numerical solution of (47.10). The degree of agreement between theory and observation will be influenced by heat losses through the thermocouple leads, so adequate measures should be taken to minimize them.

....

REFERENCES

1. M.P. Silverman and C.R. Silverman, Cool in the Kitchen: Radiation, conduction, and the Newton "hot block" experiment, *Phys. Teach.* **38**, 82 (2000).

48 Cooling II

Introduction 248
Theoretical ideas 248
Goals and possibilities 249
Experimental suggestions 249

Introduction

Cooling curves of bodies do not necessarily follow the so-called Newton's law of exponential decrease of temperature with time (see the previous project), in part because the radiation, convection, or conduction channels of heat transfers may be dominant over different temperature regimes. The following project on cooling will enable the researcher to see these effects, to evaluate the relative importance of radiation and convection, and to get a first-hand experience of designing a reliable environment and instrumentation with which to carry out heat transfer measurements.

Theoretical ideas

The basic theoretical ideas were summarized in (47.1) and (47.7) in the previous chapter. Thus the rate of heat transfer Q_C in natural convection from a body at temperature T surrounded by air at temperature T_a may be written as

$$Q_C = h(T - T_a), \tag{48.1}$$

where h is the heat transfer coefficient per unit area. For a body of finite extension situated in a convective atmosphere, h may vary along the length of the body and along its various surfaces, so it may be sensible to speak only of an average value h_{av}, itself a temperature-dependent quantity.

The corresponding quantity Q_R in radiative heat transfer is given by the Stefan–Boltzmann law, per unit area:

$$Q_R = \varepsilon \sigma \left(T^4 - T_S^4 \right), \tag{48.2}$$

where ε, σ, and T_S are respectively the emissivity of the body, the Stefan–Boltzmann constant, and some average ambient temperature. The emissivity is somewhat temperature dependent. At low temperatures (100–200°C) this dependence is very weak and it is often assumed to be a constant; in that regime its value is one for an ideal black body, and close to that for bodies painted black. However, for very shiny objects it is of the order 0.1 or less.

Let the cooling body have a total surface area A, mass M, and specific heat C. Then

$$MCdT/dt = -\int_A (Q_R + Q_C)dA. \qquad (48.3)$$

In experiments, T is measured as a function of time, as well as dT/dt as a function of both time and temperature, enabling one to deduce h_{av} by combining (48.1)–(48.3).

Goals and possibilities

- Design a draft-free enclosure such that a body in it may cool under natural convection. (Take the enclosure's surface temperature as the ambient temperature, and ensure that it remains constant over the course of each test.)
- Design a (small) draft-free enclosure, *evacuated* so that heat loss by a cooling body is by radiation only. Thermal conductivity doesn't decrease much until the pressure is quite low and the mean free path becomes limited by the dimensions of the enclosure, at which point it becomes a linear function of pressure. In this regime, it is reduced by a factor of 10 when the pressure is reduced from 1 bar to 0.3 bar [1]. Compare cooling of the same body with and without evacuating the enclosure, so as to be able to compare the importance of radiation and convection in some temperature range.
- Select appropriate thermometers like thermocouples with fast response times and low heat capacities to monitor the temperature of the cooling body and that of its surroundings. Because of good heat conduction, a small metal body can be assigned a single, reasonably uniform temperature at any time. Choose a body with an easily computed surface area.
- Consider all the requirements for amplifying thermoelectric EMFs from a thermocouple, or stable currents for a thermistor if these are used, including ways to filter out noisy signals from the amplifiers.
- Choose identical bodies with different emissivities (dark or shiny) and explore radiation effects in cooling.
- Extract from the $T(t)$ cooling curves the temperature-dependent heat transfer coefficient for the body in question.

Experimental suggestions

We suggest two different experimental setups, described in Refs [2,3], in which to investigate the contributions of convection and radiation contributions to heat loss.

Setup 1

Aluminum rods freshly turned on a lathe will have the oxides removed and a low emissivity. Identical rods painted black will have high values of ε. Use both, in order to explore the relative importance of convection and radiation.

A vertically mounted rod with a small hole drilled into it can accommodate a thermocouple to monitor its temperature, the reference end being kept at 0°C in an ice–water mixture far away. When building an enclosure, make sure that its typical lateral dimensions are an order of magnitude larger than the rod diameter. The enclosure can be made from thin sheet metal rolled into a cylinder, given a perforated lid to allow free convection currents, and supported with a gap at the bottom between it and the bench, as shown in Fig. 48.1. (Alternatively, use a large-diameter metal pipe.) A second thermometer measures T_a for the enclosure. This thermometer should be adequately shielded from radiation from the hot aluminum rod, yet able to sense the temperature of the air around the rod. The output of both thermometers may be conveniently passed on to a data acquisition system. Filter out any amplifier noise with a low-pass filter.

Figure 48.2(a), reproduced from [2], shows cooling curve data for shiny and black cylinders. A semi-log plot of the same data in Fig. 48.2(b) has curvature,

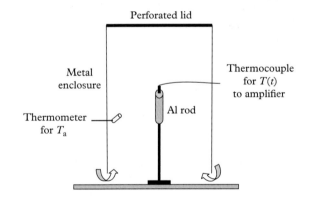

Figure 48.1 *Experimental setup 1.*

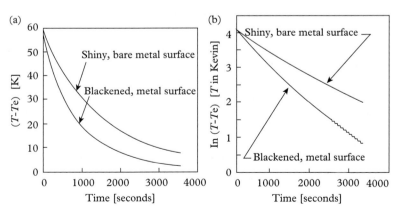

Figure 48.2 *(a) Cooling curves for shiny and black cylinders. (b) Log plot of cooling curves.*

showing that the time dependence is not exponential. In the graph, T is the temperature of the cooling object's surface and T_e is the temperature of the surroundings.

Setup 2

Rather than have having two identical bodies with different emissivities, as in setup 1, evacuate an enclosure so that the dominant mechanism for the cooling of a body is radiative loss. This cooling may then be compared with the cooling of the same body without the enclosure, or when it is not evacuated. Cooling under *forced* convection (using a pump or fan, or some such) may also be investigated in this setup [3].

The body should be small and have an easily calculable surface area. One possibility is a wire-wound resistance in a parallelepiped ceramic casing ($7\times7\times30$ mm) with a rating of 10 Ω, 15 W. If you connect it to a 12 V, 1 A supply and a rheostat, you can vary the current and with it the initial temperature. A simple solution for an evacuated enclosure is a bell jar connected to a vacuum pump. This setup is found in many schools where it is used to demonstrate absence of sound propagation in a vacuum. Once again, measure the temperature as a function of time $T(t)$ with a thermocouple connected to a data acquisition system sampling at a suitable rate.

The heat capacity C of the body is found by measuring $(dT/dt)_0$ initially when the current is switched on: $C(dT/dt)_0 = I^2 R$. Heat the body to some initial temperature (say 200°C), remove the connecting wires, and commence sampling once a steady state is reached. In vacuum, we have, from (47.1), $C(dT/dt) = \varepsilon \sigma A(T^4 - T_S^4)$.

Unlike in the previous project, extract from the T versus t data the dependence of (dT/dt) on T and verify (48.2) by plotting (dT/dt) versus T^4. The quantity T_S, the radiation temperature, may be determined by pointing an infrared thermometer at various objects in the surroundings of the cooling body and taking the average. The result may be compared with that obtained from a graphical analysis of the data.

With the bell jar removed, or with air allowed to enter the system, cooling may be monitored under conditions for which $C(dT/dt) = hA(T - T_a) + \varepsilon \sigma A(T^4 - T_S^4)$ holds. Plot (dT/dt) versus T for this case also, and compare the result with that obtained under vacuum conditions to determine the relative contributions of convection and radiation to the heat loss in the temperature range under study. By using a fan, you can also investigate cooling under forced convection. Figure 48.3, reproduced from [3], shows cooling curves under various conditions.

Figure 48.4 is a graph of dT/dt as a function of temperature T.

Figure 48.5 shows the percentage loss of heat due to radiation for natural and forced cooling.

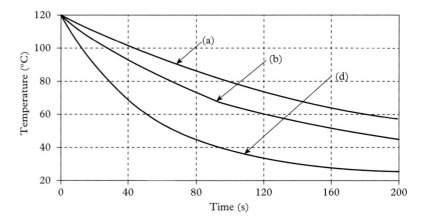

Figure 48.3 *Temperature–time curves: (a) in a vacuum; (b) air-filled bell jar; (d) forced cooling by fan.*

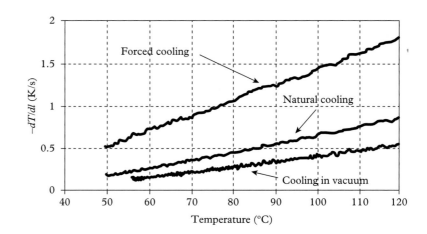

Figure 48.4 *dT/dt versus T.*

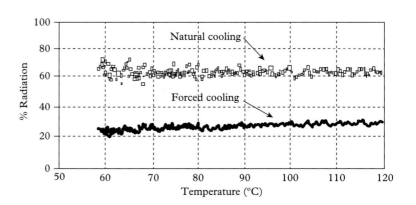

Figure 48.5 *Percentage heat loss due to radiation.*

REFERENCES

1. C.J.M. Lasance, The thermal conductivity of air at reduced pressures and length scales, available online at <http://www.electronics-cooling.com/2002/11>.
2. J.E. Spuller and R.W. Cobb, Cooling of a vertical cylinder by natural convection, *Am. J. Phys.* **61**, 568 (1993).
3. P. Twomey, C. O'Sullivan, and J. Riordan, An experimental investigation of the role of radiation in bench-top experiments in thermal physics, *Eur. J. Phys.* **30**, 559 (2009).

49 The Leidenfrost effect I

Introduction 254
Theoretical ideas 254
Goals and possibilities 255
Experimental suggestions 256

Introduction

This project is concerned with the behavior of liquids and solids in contact, but separated by a thin vapor layer which is a poor conductor of heat. The effect is familiar to anyone who has snuffed out candles with wet fingers. It has also given rise to several stunts, like dipping a hand in molten lead, or walking on hot coals. Related phenomena that may be systematically studied include drops of water dancing and oscillating on a hot skillet, and the cooling of a metallic object immersed into water or into liquid nitrogen.

Theoretical ideas

Good reviews and discussions may be found in Refs [1–3] by Walker. For a hot body immersed in a liquid, or a drop of liquid on a hot surface, an important quantity influencing the behavior is the temperature T_s, sometimes called the saturation temperature, at which the fluid will boil at the prevailing pressure.

Consider an electrically heated solid immersed in a liquid, or a metal pan of water heated from below. Suppose that we measure the power input $\dot{Q}$ per unit area A of the body needed to maintain the body at a temperature T. It is found experimentally that the rate of heat transfer in J/s depends on the temperature difference $\Delta T = T - T_s$ in the way shown in Figure 49.1 [3].

There are several regimes, characterized by different processes [3]. Initially, isolated vapor bubbles form by evaporation of the water, expand upwards by convection, and collapse to produce a sound. At higher temperatures nucleate boiling occurs. Nucleation sites enhance the bubble formation and the vapor bubbles coalesce to form columns of vapor. As the temperature is further increased above T_s, the system reaches the so-called transition boiling stage. Here, the amount of energy transferred to the water actually decreases, because the metal or pan surface is covered by a layer of vapor, which is a poor conductor of heat. This effect increases with ΔT and it is a stage that should be avoided in an efficient heat exchanger device. As the temperature is increased further, heat is transferred to the liquid above the vapor layer by radiation. This stage is labeled "film boiling" in Fig. 49.1.

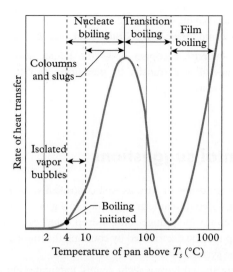

Figure 49.1 *Various regimes of heat transfer.*

Suppose that the heated body is a piece of metal which contains n moles, has a surface area A and a molar heat capacity at constant pressure C_p. Then the rate of heat loss per unit surface area of the sample is given by

$$\frac{\dot{Q}}{A} = \frac{n}{A} C_p(T) \dot{T}, \tag{49.1}$$

where the dot denotes a time derivative.

Theoretical calculations of $C_p(T)$ assume lattice models ranging from simple to more realistic. For the present experiment $C_p(T)$ may be fitted by the parameters of the Einstein or the Debye lattice model. Monitoring the temperature T and its time variation dT/dt enables one to obtain curves of $\dot{Q}$ versus ΔT.

Goals and possibilities

- Measure the lifetime of water drops of a particular size on a hot plate as a function of the temperature of the plate. Check whether the lifetime depends on the size of the drops.
- Consider the difference between distilled and tap water, and try other non-volatile liquids, such as vinegar, with different boiling points.
- Study the stability and lifetime of the drops as a function of the roughness of the hot plate.
- If you can obtain tiny metal spheres of different sizes, heat them and sprinkle them on water in a vessel. Measure the time over which these spheres glide on the vapor layer between them and the water, before the vapor evaporates.

- Water droplets having diameters in the range of 1–2 cm may perform radial oscillations around their edges. Study these (see the following chapter).
- Immerse a cylindrical copper body into liquid nitrogen and follow the various stages of heat transfer by measuring its rate of cooling and its temperature, as in (49.1).

Experimental suggestions

See Refs [3,4]. Form drops of various well-controlled sizes at the end of hypodermic needles of different diameters, attached to a glass syringe. A method that works well is to depress the plunger until a drop detaches itself from the needle. The size of a drop may be determined by counting the number needed to reduce the contents of the syringe by 1 ml.

The metal plate should have a slight central indentation in order to hold the drop falling on it in one position. It may be heated on a Bunsen or a gas range.

Thermocouples suitably calibrated are the ideal thermometer for temperatures ranging from 120°C to 400°C. Make sure that the temperature of the hot plate is stabilized. A possible setup for measuring the lifetime of a drop is shown in Fig. 49.2 [3].

For the cooling experiment in liquid nitrogen choose a massive cylindrical sample, in order to lengthen the cooling-down time. This could be a copper cylinder 5 cm in diameter, 8 cm long, weighing 1.4 kg. Copper is preferable, as it has a large thermal conductivity, ensuring that its temperature, as measured by a chromel–alumel thermocouple inserted into a hole drilled into it, may be assumed to be uniform.

It would be interesting [5] to cover the immersed metal with a thin layer of grease, a heat insulating material, in order to study the effect of different layer thicknesses, and to compare the cooling period with the bare metal's cooling

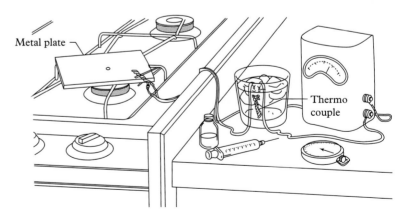

Figure 49.2 *Experimental apparatus used by Walker [3].*

curve. The cooling period may be drastically altered, and even shortened, since the presence of the insulating layer may alter the relevant heat transfer regime in the graph of Fig. 49.1.

....

REFERENCES

1. J. Walker, Drops of water dance on a hot skillet and the experimenter walks on hot coals, *Sci. Am.* **237**, 126 (1977).
2. F.L. Curzon, The Leidenfrost phenomenon, *Am. J. Phys.* **46**, 825 (1978).
3. J. Walker, Boiling and the Leidenfrost effect, available online at <http://www.wiley.com/college/phy/halliday320005/pdf/leidenfrost_essay.pdf>.
4. T.W. Listerman, T.A. Boshinski, and L.F. Knese, Cooling by immersion in liquid nitrogen, *Am. J. Phys.* **54**, 554 (1986).
5. G.G. Lavalle, P. Carrica, and M. Jaime, A boiling heat transfer paradox, *Am. J. Phys.* **60**, 593 (1992).

50 The Leidenfrost effect II: drop oscillations

Introduction 258
Experimental suggestions 258

Introduction

This project is a continuation of the previous one. Drops on a hot plate sometimes exhibit modes of vibration at temperatures well above the Leidenfrost point. The frequencies of these vibrational modes depend on the diameter of the drop as well as on the mode number. The latter characterizes the number of radial lobes in the oscillating drop. The normal mode frequencies of a freely suspended spherical water drop undergoing small oscillations have been derived by Rayleigh [1,2] and are given by

$$f_n^2 = \frac{(n-1)n(n+2)\sigma}{4\pi^2 R^3 \rho}, \tag{50.1}$$

where σ, ρ, R are the surface tension, density, and radius of the drop, respectively, and n is the node number (the degree of spherical harmonic). This formula is only a first guide, since the physical conditions on a hot plate differ from those of a freely suspended drop. In addition to the dependence of f on the ratio $\sigma/R^3\rho$ in (50.1), we would expect that on a plate f would also depend on gravity g. Dimensional analysis shows that for very small drop radii $R \ll 10^{-2}$ we may ignore gravity. See also Ref. [3].

Experimental suggestions

Control the temperature of a hot plate with a variable autotransformer. Place a copper plate on the hot plate, with a dimple at its center to contain the drop. Measure temperatures with a suitable thermocouple (iron–constantan) placed near the dimple. Use water and ice as the zero-temperature reference and convert the junction voltage to temperature using standard tables [4]. Normal mode vibrations occur at temperatures that are several hundred degrees above the boiling point.

Produce different sizes of the drops by using a syringe and suitable hypodermic needles. Measure the drop size by dropping and counting a large number of drops and measuring the change in liquid volume in the syringe. Measure drop diameter by using a traveling microscope.

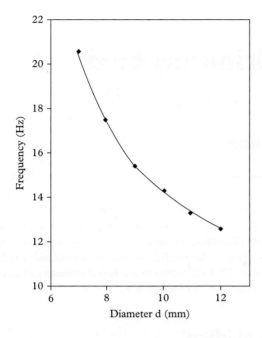

Figure 50.1 *Frequency as a function of drop diameter – continuous line is fit to $f = const.d^{-10.4}$.*

Drops may often be excited by shaking or pounding the table on which the hot plate is resting. Measure vibration frequencies using a stroboscope.

The simple experimental method described above is the basis for the data in Fig. 50.1, where the frequency of vibration is plotted as a function of drop radius, the frequency decreasing monotonically. The data can be fitted to a power law dependence, $f \propto d^n$, with $n = -1.04 \pm 0.35$. There are large uncertainties both in the measured diameter of the drops (± 1 mm) and the frequencies of vibration (± 2 Hz).

....

REFERENCES

1. J.W.S. Rayleigh, *The Theory of Sound* (Dover, New York, 1945).
2. J.W.S. Rayleigh, The influence of electricity on colliding water drops, *Proc. Roy. Soc. Ser. A* **28**, 406 (1879).
3. N. Holter and W. Glasscock, Vibrations of evaporating liquid drops, *J. Ac. Soc. Am.* **24**, 682 (1952).
4. W.M. Haynes (ed.), *CRC Handbook of Chemistry and Physics* (CRC Press, Boca Raton, FL, 2009), or later editions.

51 The drinking bird

- Introduction — 260
- Theoretical ideas — 260
- Goals and possibilities — 262
- Experimental suggestions — 263

Introduction

The drinking bird toy fascinates because it maintains a periodic motion for hours, without an apparent source of energy. It is an ideal object for an extended project, combining as it does dynamics, it being a compound pendulum of variable mass, and thermodynamics, since it is a heat engine that runs on a heat flux created by a temperature difference. There is scope here for designing sensors and data acquisition of varying sophistication in order to measure mechanical and thermal parameters, as well as for simulation to describe quantitatively both the dynamical and thermodynamic aspects of the device.

Theoretical ideas

A dyed volatile liquid such as dichloromethane (CH_2Cl_2) partially fills a tube fitted with glass bulbs at both ends. The lower end of the tube dips into the liquid in the bottom bulb (called the "body"). The upper bulb, serving as "head," holds a "beak" which serves two functions. First, it shifts the center of mass forward. Secondly, when it dips into an external beaker of liquid (usually water), its felt covering soaks up some of the liquid. As the moisture in the felt evaporates it cools the top bulb, and some of the vapor within it condenses, thereby reducing the vapor pressure of the internal liquid below that in the bottom bulb. As a result, liquid is forced upward into the head, moving the center of mass forward. The top-heavy bird tips forward and the beak dips into the beaker. As the bird tips forward, the bottom end of the tube rises above the liquid surface in the bulb; vapor can bubble up from the bottom end of the tube to the top, displacing some liquid in the head, making it flow back to the bottom. The weight of the liquid in the bulb will restore the device to the vertical position, and the cycle is maintained as long as there is water to wet the beak. The cycle is shown in the sequence of sketches in Fig. 51.1, in which the second one from the left represents the position in the photo. The line marked on the stem shows the liquid level; in the fourth sketch from the left the liquid is flowing back from the head to the bulb.

The rate of evaporation from the beak depends on the temperature and humidity of the surroundings. These parameters will clearly influence the period of the motion. Forced convection will strongly enhance the evaporation and affect

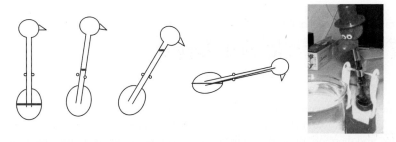

Figure 51.1 *The motion cycle of the bird.*

the period. Such enhancement will also be created by the air flow caused by the swinging motion of the bird.

The temperatures of both upper and lower bulbs must be related to the angular position of the toy. The liquid within is warmed and cooled in each cycle, but this is not the heat which operates the device. Rather, the working substance of the heat engine is the vapor of the liquid.

In any one cycle let a mass Δm_e of external liquid be evaporated, resulting in a temperature decrease ΔT of the upper bulb vapor. The two are related by [1]

$$C\Delta T = -\Delta m_e \Delta h_e, \tag{51.1}$$

where C and Δh_e are respectively the effective heat capacity of the head and the latent heat of transformation of the external liquid. The temperature decrease ΔT is accompanied by a change in vapor pressure ΔP, given by the Clausius–Clapeyron equation [2]

$$\Delta P/\Delta T = \Delta h_i/T\Delta V, \tag{51.2}$$

where Δh_i is the molar vaporization enthalpy of the working internal fluid [1]. Assuming that the vapor within obeys the (molar) ideal gas equation $PV = RT$, one finds:

$$\Delta T = \Delta P \cdot RT_A^2/(\Delta h_i P_i). \tag{51.3}$$

Here, P_i is the vapor pressure at the ambient room temperature T_A, obtained by integrating the Clausius–Clapeyron equation and is given by [2]

$$P_i(T_A) = P_0 \exp\left[-\frac{\Delta h_i}{R}\left(\frac{T_0 - T_A}{T_0 T_A}\right)\right], \tag{51.4}$$

T_0 being the boiling point of the volatile liquid, while P_0 is the normal atmospheric pressure. The pressure decrease ΔP in the head in the course of one period may be found by measuring the decrease in the internal liquid level Δy, where y is the height of the liquid column in the connecting tube above the liquid level in the lower bulb, so that the decrease is given by

$$\Delta P = \rho g \Delta y, \tag{51.5}$$

where ρ is the density of the liquid.

The period of motion τ may be calculated in terms of the rate of mass evaporation of the external liquid dm_e/dt (assumed constant for small time intervals) and the quantity evaporated Δm_e:

$$\tau = \Delta m_e / dm_e/dt = c\Delta T/(\Delta h_e \cdot dm_e/dt). \tag{51.6}$$

The inverse relationship of τ and dm_e/dt can be tested experimentally.

When the external liquid into which the bird dips is water, the evaporation rate, and therefore the period, will be dependent on the ambient humidity. In the absence of forced convection, Fick's law [3] gives the rate of evaporation in terms of the relative humidity H,

$$dm_e/dt = -\kappa(100 - H), \tag{51.7}$$

where κ is some constant. But since the oscillatory motion of the device generates convection, the above relation is expected to be replaced by $dm_e/dt = -\kappa(100 - H)^\beta$. Establishing the value of β for water is part of the project. This power law dependence would imply, via (51.6), a humidity-dependent period

$$\tau = b(100 - H)^{-\beta}, \tag{51.8}$$

b being some constant.

Finally, the device could also operate without evaporation, by creating a temperature and therefore pressure gradient between the top and the bottom, say by painting the lower bulb black and irradiating it with an electric bulb such as a floodlight from various distances [3,4].

Goals and possibilities

- Measure the period of motion of the bird and the evaporation rate and relate the two to each other. Do this for both water and several other liquids like chloroform, ethyl acetate, alcohol, and methyl alcohol as the external liquids.
- Measure the period of oscillation as a function of time.
- Establish experimentally the time range for which the evaporation may be taken as constant.
- When the external liquid is water, devise a way to change the humidity in a controlled manner in a closed environment and measure the period of motion as a function of humidity.
- For a given humidity follow the dynamics of the toy, namely its angle of inclination as a function of time, $\theta(t)$.

- By carefully measuring all the geometrical parameters of the bird, calculate its moment of inertia, then develop a dynamical model of its motion, leading to a comparison with the experimentally measured $\theta(t)$ data.
- The toy is a heat engine operating between two heat reservoirs. It is therefore of interest to calculate its efficiency. To do this one must calculate the mechanical work done on raising the internal liquid column to the top, before it is allowed to drain back. This in turn requires locating experimentally the center of gravity of the bird at the beginning of the cycle when the column begins to rise, and when the column is at its maximum height.
- Monitor the temperature variations of the head and the tube by means of tiny thermocouples attached to them, and try to understand the thermodynamics and heat transfer occurring during the motion. Very small thermistors might also do, but care must be exercised in wiring; use the finest enamel wire about 0.1 mm in diameter.
- Carry out the investigation when the temperature gradient is produced by irradiating the blackened lower part of the toy with a lamp, while shielding the head. This causes the same periodic dipping phenomenon as that caused by evaporation from the head.

Experimental suggestions

To measure evaporation rates the bird may be placed on a sensitive electronic balance, accurate to 0.001 g. A few drops of various external liquids may be applied to the felt of the head by a pipette. Measure the time variation of the mass of this liquid, and that of the period of motion, without replenishing the liquid when the bird bows into its horizontal position. Allow for the time spent in the horizontal position.

The effects of humidity may be observed if the toy is enclosed in a transparent container which allows one to vary the humidity.

The effective heat capacity of the head C may be calculated from the temperature change in one period ΔT from (51.1), together with (51.3) and (51.5).

Since the toy is very cheap, one may be dismantled in order to measure all the geometrical parameters needed for a dynamical simulation: the external and internal diameters of the tube and bulbs, and the masses of the beak and the hat.

The simulation for the time dependence of the angle of inclination $\theta(t)$ will need the moment of inertia and the torque relative to the rotation axis of the toy, as functions of the level height of the internal liquid. Calculating these is in itself a valuable exercise in rotational dynamics [5,6]. Friction is present and affects the motion: it could be modeled by adding to the restoring torque a term that is proportional to the angular velocity, the constant of proportionality being an adjustable parameter chosen to fit the damping of the motion.

To monitor the angular position and the time dependence of the dipping interval, data that could be compared with the dynamic simulation, you could use following method (described in detail in [5]).

1. Mount a powerful small magnet to the underside of the bird's body, with its polar axis parallel to the tube. Its mass must be balanced on the top of the head by putty, in order not to affect the pivoting.
2. Mount a tiny sensor coil, resistance and inductance of the order of 400 Ω and 6 mH, below the bottom bulb, its very thin (0.1 mm diameter) wires connected across a resistance of some 300 Ω.
3. Use adhesive putty to position a small single-axis magnetometer sensor some 20 cm away from the base of the bird. The small magnet generates a field B along the sensing axis of the magnetometer which will vary monotonically with the angle of inclination of the toy. The sensing device may be calibrated by using a protractor to position the bird at a series of accurately measured angles θ, and using a suitable polynomial curve to fit the resulting $B(\theta)$ data.
4. From thin (0.12 mm) chromel and alumel wires make thermocouples, suitably calibrated, and glue them on the head's felt and on the tube. These can then be used to follow the time dependence of the head and stem temperatures in each cycle, correlating these with the angular motion of the bird. Heat enters the tube and is transported to the head, and this will be reflected in a steady state temperature difference between the two. Both head and tube temperatures may vary during a cycle, and these variations can then be related to heat transfer from the surroundings and evaporation enhancement due to the convection generated by the swinging motion.

An alternative would be to use a glass-encapsulated bead thermistor the size of a rice grain (10 kΩ at 20°C), suitably connected with very thin wire (38 AWG) to a power source and data logger. But its response time would be too slow, since its thermal mass is much larger than the thermocouple, and so it would monitor the changes in temperature during the swings in less detail. Perhaps smaller thermistors are available.

Paint the lower bulb and tube black, and irradiate them with a flood lamp at controlled distances, while shielding the head, so as to create a temperature gradient between head and body. Such heating increases the vapor pressure within, causing liquid to be forced up into the head and making the toy dip, just as for the cooling of the head by evaporation. It will then be an interesting exercise to relate both the time elapsed before the first swing and the period of motion to the following parameters: the effective surface being illuminated; the power input of the lamp ($I \cdot V$); the effective energy supplied, which is dependent on the lamp's distance from the bird (inverse square law); and the parameters of the internal liquid (C, Δh_i, and ρ).

REFERENCES

1. J. Guemez, R. Valiente, C. Fiolhais, and M. Fiolhais, Experiments with the drinking bird, *Am. J. Phys.* **71**, 1257 (2003).
2. F. Reif, *Fundamentals of Statistical and Thermal Physics* (McGraw-Hill, New York, 1965).
3. J. Guemez, R. Valiente, C. Fiolhais, and M. Fiolhais, Experiments with the sunbird, *Am. J. Phys.* **71**, 1264 (2003).
4. R. Mentzer, The drinking bird – the little heat engine that could, *The Physics Teacher* **31**, 126 (1993).
5. R. Lorenz, Finite-time thermodynamics of an instrumented drinking bird toy, *Am. J. Phys.* **74**, 677 (2006).
6. L.M. Ng and Y.S. Ng, The thermodynamics of the drinking bird toy, *Phys. Educ.* **28**, 320 (1993).

52 Liquid–vapor equilibrium

Introduction 266
Theoretical ideas 266
Goals and possibilities 267
Experimental suggestions 267

Introduction

The liquid–vapor line in the phase diagram is described by the Clausius–Clapeyron equation. The present project shows how to obtain the pressure versus temperature curve of a liquid using simple apparatus, relate the measurements to an often-used empirical formula, and extract the mean latent heat of vaporization in a limited temperature range. While the apparatus is designed for water, other liquids could be investigated.

Theoretical ideas

The slope of the liquid–vapor coexistence curve is given by [1]

$$\frac{dP}{dT} = \frac{L_{12}}{T\Delta V}, \tag{52.1}$$

where L_{12} is the latent heat, $\Delta V = V_2 - V_1$ is the change in volume between the vapor phase 2 and the liquid phase 1, and P, T are the pressure and absolute temperature, respectively. Since $V_2 \gg V_1$, and if the vapor obeys the ideal gas equation for n moles,

$$PV_2 = nRT, \tag{52.2}$$

then $\Delta V = nRT/P$. If L_{12} is assumed to be temperature independent, (52.1) can be integrated to give

$$P = P_0 \exp(-L_{12}/nRT), \tag{52.3}$$

where P_0 is a constant. The vapor pressure is a rapidly increasing function of temperature, the rate being governed by the latent heat of vaporization.

For water, a useful, purely empirical, relationship between the vapor pressure and temperature is the Antoine equation [2,3]:

$$P(\text{kPa}) = \exp\left[16.573 - \frac{3988.842}{T - 39.47}\right]. \tag{52.4}$$

Similar equations are available for ethanol and a whole series of organic compounds [2].

Liquid–vapor equilibrium

Goals and possibilities

The goals are to determine experimentally the vapor pressure versus temperature curve and to test the quality of this data by comparing with the predictions of the Antoine equation available in the literature. Water and ethanol are recommended, but other substances could be tried.

Experimental suggestions

Here we describe the experimental setup for one of the many possible procedures for producing the P–T curve [4]. With this particular setup you can take into account the expansion of the liquid as the temperature is raised.

The apparatus required is: glass tubes with internal diameter $d = 0.6$ cm and lengths 10, 15, and 150 cm; two cylindrical heat-resistant jars of volume about 150 cm^3 and internal diameter $D = 5$–6 cm fitted with rubber stoppers; a larger vessel and ice; hot plate and magnetic stirrer for heating; digital thermometers or thermocouples; degassed liquids to be investigated (distilled water, ethanol, etc.).

Figure 52.1 shows jars A and B fitted with rubber stoppers. In the initial state, Fig. 52.1(a), jar A is filled with the liquid and a 15 cm-long glass tube is fitted, together with the thermometer, through the stopper. Air bubbles must be eliminated. Jar B is two-thirds filled with the liquid and likewise fitted, through the stopper, with a thermometer and a 10 cm tube reaching almost to the bottom. The system is cooled with ice, the lowest temperature is measured, then liquid is added or removed from both jars with a pipette until its level in the tubes just reaches the upper level of the rubber stopper. Jar A will then contain

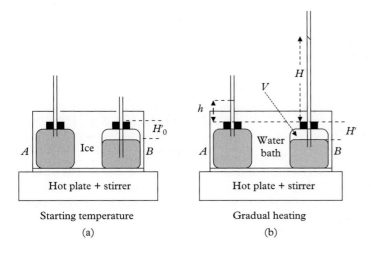

Figure 52.1 *Experimental setup for vapor pressure as function of temperature.*

only liquid, B both liquid and a mixture of air and vapor. The volumes of liquid and gas can be found by weighing the vessels and tubes with and without the liquids.

By Dalton's law of partial pressures, the total gas pressure P_B in B is the sum of the vapor pressure P_{vi} and the air pressure P_a, $P_B = P_{vi} + P_a$, and also $P_{vi} + P_a = P_{atm} + \rho_i g H'_0$, where P_{atm} is the atmospheric pressure and ρ_i is the liquid density at the initial temperature.

In the heating stage, Fig. 52.1(b), the glass tube in B is extended by some 1.5 m, to allow for the rise of the water level with the heating as the temperature and gas pressure rise. Some means must be arranged to measure accurately the heights H and h of the liquid levels in the tubes attached to both vessels, and the height H' of the gas space in B. Controlled heating should be slow, allowing stability to be attained at intermediate temperatures.

The calculation of the vapor pressure is simplified if the following assumptions are made:

1. The water density is taken to be constant in the 280–360 K temperature range (introducing an error less than 3%).
2. The difference $P_{vi} - \rho_i g H'_0$ may be neglected compared to P_{atm} (near the ice point P_v is only 0.6 kPa).
3. The temperature of the liquid in the long tube is that of the surroundings and the expansion of the liquid in the tube itself is negligible.

The initial volume V_i and the volume V of the gas in B at any temperature are related by

$$V = V_i + \frac{\rho}{\rho_i} SH - \frac{m_{Bi} - \rho SH}{m_{Ai} - \rho Sh} \frac{\rho}{\rho_i} Sh. \qquad (52.5)$$

Here, $S = \pi d^2/4$, m_{Ai} and m_{Bi} are the masses of the liquids in A and B, including those in the tube within. The vapor pressure at any temperature is given by

$$P_v = P_{atm}\left(1 - \frac{V_i T}{V T_i}\right) + \rho_i g \left\{H'_0 + \left[1 + \frac{S}{S'}\right]H\right\}. \qquad (52.6)$$

Here, $S' = \pi D^2/4$ is the cross-sectional area of the jars.

Figure 52.2, reproduced from [3], shows typical data for $\ln(P_V)$ as function of $1/T$, and a comparison with the Antoine formula (52.4). The most important contribution to errors in the data on which the graph is based is the measurement of volume. This could be reduced by using a cathetometer, but these are not available in most schools or colleges.

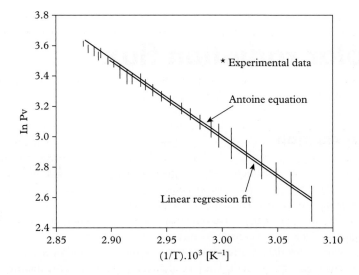

Figure 52.2 *Experimental data compared to empirical formula, for $ln(P_v)$ as a function of $1/T$.*

REFERENCES

1. F. Reif, *Fundamentals of Statistical and Thermal Physics* (McGraw-Hill, New York, 1965).
2. Wikipedia, Antoine equation, online at <http://en.wikipedia.org/wiki/Antoine_equation>.
3. S. Gesari, B. Irigoyen, and A. Juan, An experiment on the liquid vapor equilibrium of water, *Am. J. Phys.* **64**, 1165 (1996).
4. S. Velasco, J. Faro, and L. Roman, An experiment for measuring the low temperature vapor line of water, *Am. J. Phys.* **68**, 1154 (2000).

53 Solar radiation flux

- Introduction — 270
- Theoretical ideas — 270
- Goals and possibilities — 271
- Experimental suggestions — 271

Introduction

The flux of solar radiation and its variation at various times of the day, and for different days and seasons of the year, is an important factor in determining the energy efficiency of solar devices, influences local climate, and is relevant in many environmental issues [1]. Many heat flux sensors are available commercially to measure the solar flux (see, for example, <http://www.industrialsensorresource.com>, <http://www.appropedia.org>, or <http://www.hukseflux.com>). The goal of this project is to measure this flux with an apparatus built by the researcher. The project is an exercise in heat transfer.

Theoretical ideas

The heart of the device to be built is a pair of parallel metal plates exposed to the sun, one painted black, the other painted white. Radiation falling on the plates causes differential heating due to their color and therefore a difference in their temperature, which is measured. This temperature difference is a measure of the solar flux at any time. This type of radiometer has to be enclosed in some enclosure with a hemispherical window in order to reduce both convection and reflection losses.

Any radiometer needs to have a response time that is short compared to the time scale over which radiation changes significantly. The experimental section will explore these thermal transients in order to gain an estimate of the response time of the device.

In a first, very approximate, approach, and in order to fix ideas, we assume that the plates have each reached a steady temperature T_b, T_w (black and white, respectively), that the absorption coefficients of the plates a_b, a_w do not depend on the wavelength of radiation, and that there is a constant ambient temperature T_s. Denote by K_d, K_v, and K_r coefficients related to the conduction, convection, and radiation processes, respectively. In general they are different for the black and white plates. If the radiant flux is U, the energy balance gives, for the black plate,

$$a_b U = K_d(T_b - T_s) + K_v(T_b - T_s)^n + K_r(T_b^4 - T_s^4), \tag{53.1}$$

and a similar equation for the white plate, where n is the exponent of the convection term, equal to 5/4 for free convection. If the temperature rise of the plates is small, say from 300 to 302 K, and if we take $n = 1$, the replacement of $T_b^4 - T_s^4$ by $T_b - T_s$ is not unreasonable, and one sees from the energy equations that

$$a_b US = K_B S(T_b - T_s), \qquad (53.2)$$

$$a_w US = K_W S(T_w - T_s). \qquad (53.3)$$

Here, K_B and K_W are new coefficients. We may therefore write for the temperature difference of the plates:

$$T_b - T_w = (a_b/K_B - a_w/K_W)U, \qquad (53.4)$$

so that it is proportional to the solar energy flux. The device to be built will make use of this relationship. A more complete, numerical analysis is needed if the assumptions leading from (53.1) to (53.4) are not valid.

To discuss the response time of a single blackened plate of mass m and specific heat c, exposed suddenly to an energy flux U, consider the temperature rise dT in a time dt. Then energy conservation gives

$$mcdT = Ua_b S dt - K_B(T_b - T_s). \qquad (53.5)$$

This equation may be solved numerically on a spreadsheet with parameters suitably chosen to fit the experimental conditions.

Goals and possibilities

- Build a thermocouple to measure temperatures.
- Study experimentally the transient response of a single blackened metal slab when exposed to a sudden energy flux.
- Build an experimental cell with a hemispherical window to house the metal plates equipped with thermocouples.
- Calibrate the output of the thermocouples in the device by comparing with available solar insolation data for the time, place, and latitude of the experiment, or with a commercial pyranometer.
- Follow the solar flux throughout the day, in various seasons and various cloud covers.

Experimental suggestions

See Ref. [2]. Thermocouples may be built from iron and constantan 0.1 mm diameter wires. The output of a thermocouple may be read on a sensitive voltmeter able to read tenths of millivolts, or by using a DC amplifier with a multimeter.

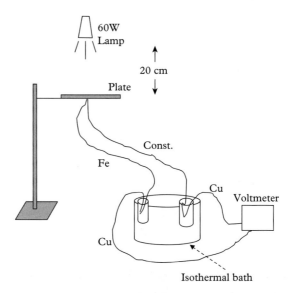

Figure 53.1 *Setup to measure time response of blackened plate.*

Use aluminum plates (say 12×80 mm), one painted matte black, the other glossy white. Affix the thermocouples to the center of the plates, using a small amount of epoxy glue (so as not to affect the thermocouple reading) to ensure good thermal contact.

Mount the plates inside an aluminum enclosure (serving as a wind shield) painted white on the outside, just behind a hole of some 8–10 cm diameter, cut on the upper side of this box facing the sun. Cover the hole with a hemispherical transparent plastic surface, through which solar radiation will fall on the two plates. Such a surface may be found on the top of a disposable cup often used to serve hot drinks in fast food stores (a poorer alternative is a watch glass from the chemistry lab, but that is not a hemisphere). Small holes on one wall of the enclosure allow the thermocouple wires to lead to the voltmeter.

Measure the time response of a blackened plate using the setup shown in Fig. 53.1, holding the two low-temperature junctions at a fixed known reference temperature in a beaker. The thermocouple output is recorded as a function of time, and the time constant, the time needed to reach 63% of the final temperature, is measured.

The time response of the slabs in the box may be measured in a similar way, giving an idea of the time resolution of the device and its ability to register time-varying insolation conditions.

....

REFERENCES

1. Wikipedia, Sunlight, online at <http://en.wikipedia.org/wiki/Sunlight>.
2. V. Zanetti and A Zecca, A home-made pyranometer for solar experiments, *Am. J. Phys.* **51**, 683 (1982).

Appendix A
Project ideas

In this appendix we suggest a number of interesting phenomena to study, without going into details at the level presented in earlier chapters. Brief descriptions are followed by references providing background information. You can use this to see whether the topics are suitable in terms of theoretical background, equipment available, time invested, and level of competence.

Drop formation and capillary waves in a falling stream of liquid

A stream of liquid issuing from a tap narrows in cross section as it falls, then breaks up into drops, as in the Fig. A.1. The mechanism and the height at which breakup occurs is a process very much influenced by the surface tension, viscosity, and density of the liquid. When the volume flux from a tap is such that the stream has a diameter of 2–3 mm, obstructing the stream with a rod several centimeters from the tap generates a stationary field of varicose capillary waves (Chapter 5 of [1]) upstream of the rod. If the rod is dipped first into a liquid detergent (lowering the surface tension) and then inserted into the stream, the capillary waves begin at some critical distance above the rod, below which the stream is cylindrical. Much theoretical discussion has been devoted to these topics [2–7] from the 19th century (Rayleigh, Plateau) to the present day. An excellent introduction to these topics is the six-lecture MIT module on surface tension by Bush, available online [1].

Quantitative experiments could measure the narrowing of the stream with height and compare the measured stream profile with that predicted by the equation of continuity. You could study the depth below the tap at which breakup into drops occurs as a function of surface tension and viscosity, and the appearance of capillary waves as a function of surface tension. You might also measure, using any of the several known methods, the surface tension and the viscosity of the liquids employed. One method for the surface tension is described in Chapter 33. You could use high-resolution and high-speed photography [8] to study the structure and time evolution of the drops.

Figure A.1

REFERENCES

1. J.W.M. Bush, MIT Lecture Notes on Surface tension (2004), available online at <http://web.mit.edu/2.21/www/Lec-notes/Surfacetension>.
2. V. Grubelnik and M. Mahrl, Drop formation in a falling stream of liquid, *Am. J. Phys.* **73**, 415 (2005).
3. J. Eggers, Nonlinear dynamics and breakup of free surface flows, *Rev. Mod. Phys.* **69**, 865 (1997).
4. X.D. Shi, M.P. Brenner, and S.R. Nagel, A cascade of structure in a drop falling from a faucet, *Science* **265**, 219 (1994).
5. M.J. Hancock and J.W.M. Bush, Fluid pipes, *J. Fluid Mech.* **466**, 285 (2005).
6. J.S. Rowlinson and B. Widom, *Molecular Theory of Capillarity* (Dover, New York, 2003).
7. P-G. de Gennes, F. Brochard-Wyart, and D. Quéré, *Capillary and Wetting Phenomena* (Springer, New York, 2000).
8. D.H. Peregrine, G. Shoker, and A. Symon, The bifurcation of liquid bridges, *J. Fluid Mech.* **212**, 25 (1990).

Gravitational surface waves in water

Consider a rectangular trough containing water to a depth h, in which gravitational surface waves are generated by a paddle oscillating at controlled frequencies ω. For small amplitudes the frequency–wavelength (dispersion) relation is given by [1,2]

$$\omega^2 = (2\pi/\lambda) g \tanh(2\pi h/\lambda),$$

where g is the acceleration due gravity.

When the amplitude becomes comparable to the depth h, the non-linearities that arise will generate subharmonics of the driver frequency [3].

As the paddle frequency is increased, edge waves will be generated at $\omega = \omega_{\text{driver}}/2$. Their wavelength is twice the width of the trough. Only the water very close to the paddle will slosh back and forth horizontally across the width of the trough. The equation governing the edge waves is a Mathieu equation [4]. The wave amplitude decays exponentially with the distance along the length of the trough, and at the far end of the trough the water is almost quiescent. Since the system resembles a waveguide, this decay may be understood in terms of acoustic waveguide theory [5].

The study of frequencies for various amplitudes, checking the validity of the dispersion relation, the detection of subharmonics, and the investigation of edge waves form the core of the project.

A possible experimental setup is shown in Fig. A.2, with a trough of dimensions $20(L) \times 2.8(W) \times 8(H)$ cm. Wetting agent is added to the water to prevent

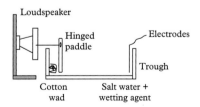

Figure A.2

the meniscus being pinned to the wall, and the water is made conducting by adding a small amount of NaCl. The cotton wad at the back of the paddle helps damp out large-amplitude oscillations. The wave detector is a pair of vertical wire detectors connected to a 10 V DC source. They are aligned parallel to the wavefront and dipped into the water to a level just above the bottom. Wave motion causes the height of the water to oscillate. This is reflected in the variation of the current and the conductance between the electrodes, which is proportional to the height of the wave. Details for the data acquisition electronics may be found in [6].

REFERENCES

1. L.D. Landau and E.M. Lifshits, *Fluid Mechanics* (Pergamon, London, 1969), p. 40.
2. Wikipedia, Dispersion (water waves), online at <http://en.wikipedia.org/wiki/Dispersion_(water_waves)>.
3. R. Keolian, L. Turkevich, S. Putterman, and I. Rudnick, Subharmonic sequences in the Faraday experiment, *Phys. Rev. Lett.* **47**, 1133 (1981).
4. C.J. Garrett, On cross waves, *J. Fluid Mech.* **41**, 837 (1970).
5. L.E. Kinsler, A.R. Frey, G. Williams, and J.V. Sanders, *Fundamentals of Acoustics* (Wiley, New York, 1982), p. 216.
6. J. Wu and I. Rudnick, An undergraduate experiment designed to study the dispersion relation of gravitational surface waves in water, *Am. J. Phys.* **52**, 1008 (1984).

Synchronization of metronomes

Interacting systems that result in synchronized motion arise in innumerable fields: biological systems [1], clapping in audiences [2], neutrino oscillations [3], chaotic systems [4], and voltage oscillations in Josephson junctions (see the list of references in [5]) are but a few examples.

The suggestion here is to study the simple system of two pendulum metronomes, coupled by placing them on a light, easily movable platform which can move horizontally on two rollers. If the frequencies of oscillations differ slightly, the pendulums will synchronize after some tens of seconds. The audible tick of the metronomes can be recorded, enabling one to follow the approach to synchronization and the evolution of the phase differences. It is interesting to explore the effect of increasing frequency difference on the phase-locked synchronized motion. Beyond some threshold difference one does not obtain synchronized motion.

The experimental parameters that can be varied and which affect the synchronization process are the average frequency, the frequency difference, and the base mass. The damping of the base platform motion can have a significant effect on the phase difference between the metronomes in the synchronized mode.

Following the base motion with a sonic ranger or a fast video camera would enable one to monitor the degree of synchronization.

A number of theoretical models for the synchronization of coupled oscillators could serve as guidelines for further theoretical analysis of the experimental data. A paper by Pantaleone [5] presents a simplified analysis based on the van der Pol equation. Several metronomes oscillating on a common rolling platform would enable you to realize mechanically a model in which a phase transition occurs from incoherent oscillations to collective synchronization as the coupling parameter (different base masses) is increased.

REFERENCES

1. H.H. Strogatz and I. Stewart, Coupled oscillators and biological synchronization, *Sci. Am.* **269**, 102 (1993).
2. Z. Neda, E. Ravasz, Y. Brechet, T. Vicsek, and A.L. Barabasi, The sound of many hands clapping, *Nature* **403**, 849 (2000).
3. J. Pantaleone, Stability of incoherence in an isotropic gas of oscillating neutrinos, *Phys. Rev.* **D58**, 073002 (1998).
4. L.M. Pecora and T.L. Caroll, Synchronization in chaotic systems, *Phys. Rev. Lett.* **64**, 821 (1990).
5. J. Pantaleone, Synchronization of metronomes, *Am. J. Phys.* **70**, 149 (2002).

Non-linear dynamics of a compass needle

A freely rotating magnetic dip needle is placed along a constant vertical magnetic field of a solenoid B_S. It is made to rotate by a sinusoidally alternating magnetic field B_H of a Helmholtz coil pair, driven by a function generator at about 1.5 Hz. The field of the Helmholtz coil can be either vertical, as in Fig. A.3, or horizontal. A photogate suitably positioned will measure the blocking time as the magnet cuts through its infrared beam. Employing a pair of photogates enables one to measure both the magnitude and direction of the angular velocity. The blocking time is inversely proportional to the angular velocity of the needle, which is the dynamical variable of this electromechanical system. Fields are adjustable so that, for example, the needle may be made to rotate in one direction. Paper baffles may be attached to the tips of the needle in order to alter the air resistance force, denoted by a damping parameter b, hindering the rotation.

By varying the relative magnitudes and directions (parallel or at right angles) of B_S and B_H and the damping parameter b, many interesting effects may be investigated with this system. Among these are its stability to perturbations, the period-doubling bifurcation route to chaos, transient behavior and its decay, strange attractors, and higher-order periodicities. Numerical solutions

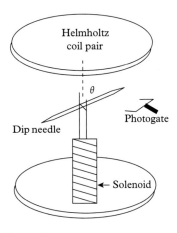

Figure A.3

of the non-linear differential equation governing the rotation of the magnetic needle having magnetic moment μ and moment of inertia I (say when B_S and B_H are parallel),

$$I\ddot{\theta} = -\mu B_S \sin\theta - b\dot{\theta} + \mu B_H \sin\theta \cos\omega t,$$

enable the modeling of the behavior of the needle as a function of the parameters b, B_S, and B_H.

....

REFERENCES

1. A. Ojcha, S. Moon, B. Hoeling, and P.B. Siegel, Measurement of the transient motion of a simple nonlinear system, *Am. J. Phys.* **59**, 614 (1991).
2. T. Mitchell and P.B. Siegel, A simple setup to observe attractors in phase space, *Am. J. Phys.* **61**, 855 (1993).

Non-linearities and the bent tuning curve

Whether in mechanics or in acoustics, resonance curves for forced harmonic systems are symmetrical about the resonance frequency and are reasonably well fitted with a Lorentzian function, as shown in a number of projects on acoustics in Part 3. By contrast, non-linear anharmonic force terms can result in bent, non-symmetrical tuning curves (amplitude versus frequency plots). To show this experimentally, and to solve by numerical methods the relevant Duffing equation modeling the phenomenon, is a worthwhile project.

Experimentally, the task of finding the bent tuning curve may be accomplished with simple equipment [1]. You need a 6 in loudspeaker, driven by a signal generator able to vary the frequency to an accuracy of 0.1 Hz, a frequency meter, an audio amplifier, and a video camera. The vibrating element can be a rubber string made from a rubber band, one end of which is attached to a rod glued to the diaphragm of the loudspeaker, the other end to a stand which one may raise or lower at will, thus varying the tension. This needs to be done carefully with a suitable mechanism, as it has been shown [2,3] that non-linear effects come into prominence when the string is closest to its relaxed length, while for larger strains one recovers the usual symmetrical resonance curve. The video camera is used to display the amplitude of vibrations on a monitor, allowing measurement, once a suitable scale is established by placing a meter ruler in the plane of vibration [1].

Theoretically, the model equation of Duffing has been shown [4] to reproduce well the hysteresis phenomenon and the deviation from the Lorentzian curve.

It includes the forcing, dissipative, and anharmonic effects represented by F, β, and α, respectively, in the equation

$$\ddot{x} + \beta\dot{x} + (1 + \alpha x^2)x = F\cos(\omega t). \tag{A.1}$$

The parameters needed for the numerical solution of (A.1) may be fitted from fits to experimental data, or calculated from the characteristics of the rubber string and the drive.

REFERENCES

1. R. Khosropour and P. Millet, Demonstrating the bent tuning curve, *Am. J. Phys.* **60**, 429 (1992).
2. J.A. Elliott, Intrinsic nonlinear effects in vibrating strings, *Am. J. Phys.* **48**, 478 (1980).
3. J.A. Elliott, Nonlinear resonance in vibrating strings, *Am. J. Phys.* **50**, 1148 (1982).
4. N.B. Trufillaro, Nonlinear and chaotic string vibrations, *Am. J. Phys.* **57**, 408 (1989).

Surface wave patterns in a vertically oscillating liquid

When a viscous fluid is made to undergo vertical oscillations, acceleration beyond a certain threshold value will result in a pattern of standing waves on its free surface. The phenomenon is known as the Faraday instability or Faraday waves. For a single driver frequency, patterns of squares or lines are generated that depend on the viscosity of the liquid. If the driver is fed with a superposition of two frequencies, hexagonal and more complex, beautiful patterns may be excited on the surface. A project devoted to investigating these effects will be an introduction to the current research on non-linear effects and pattern formation. The important parameters to be varied will be the driving frequency, the amplitude of forcing, the viscosity of the liquid, and the mixing phase angle in the case of a superposition of two frequencies.

The equipment required is the following [1]: A 6″ loudspeaker, a 6 W audio amplifier, and a function generator that may be programmed from a computer to generate single-frequency sinusoidal signals, or a mix of two frequencies (see Fig. A.4), a rigid light dish (made from Kevlar; suitable dimensions are 8 cm in diameter and 1 cm deep – see Fig. A.5) that may be stuck to the cone of the speaker, and a bubble spirit level for leveling and balancing the system. Silicone oil as the liquid gives excellent results. Experimental and theoretical papers [2–4] should be consulted before embarking on the project.

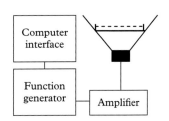

Figure A.4

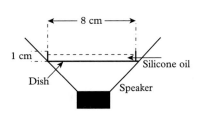

Figure A.5

REFERENCES

1. T. Pritchett and J.K. Kim, A low-cost apparatus for the production of surface wave patterns in a vertically oscillating fluid, *Am. J. Phys.* **66**, 830 (1998).
2. F. Melo, P. Umbanhower, and H.L. Swinney, Transition to parametric wave patterns in a vertically oscillated granular layer, *Phys. Rev. Lett.* **72**, 172 (1994).
3. F. Melo, P. Umbanhower, and H.L. Swinney, Hexagons, kinks, and disorder in oscillated granular layers, *Phys. Rev. Lett.* **75**, 3838 (1995).
4. S.T. Milner, Square patterns and secondary instabilities in driven capillary waves, *J. Fluid Mech.* **225**, 81 (1991).

Hexagonal pencil rolling on an incline

The motion of a cylinder with hexagonal cross section down an inclined plane consists of a series of inelastic collisions between the faces of the cylinder and the plane. Kinetic energy is lost, in an amount which depends on the coefficient of restitution. The body will acquire a terminal velocity. Each of the six edges serves as an instantaneous axis of rotation.

The system lends itself to experimental analysis [1] by filming its motion, and by recording with a sensitive microphone the sound made by the impact of the cylinder as it rolls. A mixture of rolling and sliding for incline angles θ such that $\mu < \tan\theta$, where μ is the coefficient of friction, makes the problem more interesting. The material of the inclined plane may be varied, affecting the coefficient of restitution and hence the terminal speed.

The theoretical analysis can be an excellent extended exercise in rotational dynamics, collisions, and rolling. It can be performed at various degrees of sophistication, including questions such as the influence of non-sharp hexagonal edges, the effective face area, calculation of the coefficient of restitution, and conditions for loss of contact between the cylinder and the incline.

REFERENCE

1. A.R. Zadeh, Motion of a hexagonal pencil on an inclined plane, *Am. J. Phys.* **77**, 400 (2009).

Granular materials

This is a subject of active research. Some examples of granular materials are nuts, coal, sand, rice, coffee, corn flakes, fertilizer, and ball bearings. Some naturally occuring phenomena with granular materials are booming sand dunes, sand ripples, and un-mixing of unlike grains. Granular materials are commercially important in applications as diverse as the pharmaceutical industry, agriculture (silo honking), and energy production. Good review articles [1–3] and a book [4] are available to guide entry into this fascinating field. The following topics may explored with simple equipment:

- The time dependence of flow from different-shaped containers.
- The flow rate as a function of orifice diameter.
- The segregation into regular patterns in a rotating cylinder containing two different granular materials.
- The angle of repose for different grains.
- The influence on the speed of flow and on the angle of repose of adding small amounts of liquids of different viscosities to the grains. This simulates the presence of humidity in silos.
- The velocity distribution of grains in a vertically vibrating container. This requires the use of high speed photography.

REFERENCES

1. H.M. Jaeger and S.R. Nagel, Physics of the granular state, *Science* **255**, 1523 (1992).
2. H.M. Jaeger, Sands, jams and jets, *Physics World* **18**, 34 (2005).
3. H.M. Jaeger, S.R. Nagel, and R.P. Behringer, Granular solids, liquids and gases, *Rev. Mod. Phys.* **68**, 1259 (1996).
4. R.A. Bagnold, *The Physics of Blown Sand and Desert Dunes* (Methuen, London, 1941).

Charging by pouring granular materials

Pour well-dried granular materials, such as quartz sand, sugar, or rice from containers made from different materials into a metal can. Design and calibrate an electrometer capable of measuring the quantity of charge produced on the can. How does the amount of charge depend on the material of the container? Using a video camera, investigate the behavior of the falling stream as the can fills up. Follow the changes when you shine a UV lamp on the stream.

Physics of the Moka espresso coffee maker

See Ref. [1]. This device, available in all sizes on the internet, is shown in Figs A.6 and A.7, and has the following parts: A boiler with a safety valve, and bottom

Figure A.6 *The Moka espresso.*

container (Fig. A.6); a container in the shape of a funnel which contains the ground coffee – it has a base pierced with holes, seen in Fig. A.7, with a spout at its end, and is lowered into the boiler, but does not touch the water in the latter; a metal filter with a rubber gasket around it, separating the funnel from the boiler; a pot, top right in Fig. A.7, where the ready coffee collects.

To prepare the coffee, water is poured into the boiler to a level just below the safety valve. Place the funnel into the boiler, fill it with ground coffee, screw on the pot and place the device on a low flame. After a few minutes the water rises, is forced through the coffee, through the funnel and into the pot. When the residual steam rises from the boiler to the pot, a gurgling sound is heard and the coffee is ready.

Temperature probes can be inserted into the safety valve and into the exit of the coffee in the pot, so as to monitor the time dependence of the temperature, and the development of steam. The device is a form of steam engine; its efficiency can be calculated by measuring the rate of heat supplied and delivered. Filtration pressure, and the filtration coefficient of the coffee, can be measured and calculated. Times from the onset of heating and the first appearance of coffee may be calculated and compared with measured values. The influence of the initial amount and temperature of water, free-space volumes, different coarseness of the coffee grind, and heating rates on the amount and final temperature of the coffee extracted can be investigated. A theoretical analysis could be attempted, based on the relation between the pressure drop across the coffee plug, its conductance (not necessarily a constant), and the volume flow rate, involving the law of Darcy [2], which is a phenomenologically derived equation describing the flow of liquid through a porous medium, relating the flux to the pressure gradient. The Clausius–Clapeyron equation can be used to predict the increase in vapor pressure on heating the water, and an ideal gas law employed to treat the partial pressure of the air trapped when the vessel is sealed.

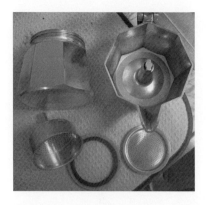

Figure A.7 *The Moka parts.*

....

REFERENCES

1. W.D. King, The physics of a stove-top espresso machine, *Am. J. Phys.* **76**, 558 (2008).
2. Wikipedia, Darcy's law, online at <http://en.wikipedia.org/wiki/Darcy's_law>.

Contact angle and surface roughness

See Refs [1,2]. A measure of surface wettability is the contact angle between a liquid drop and the solid surface: hydrophobic/hydrophilic surfaces have contact angles greater/less than 90°. Since only a fraction of the droplet may be in contact with a rough surface, the contact angle may be significantly affected by the degree of roughness.

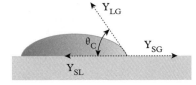

Figure A.8

The contact angles defined by Young and shown in Fig. A.8 may be determined on smooth sheets of plastic like Teflon or plexiglass, cleaned with ethanol and distilled water. Other surfaces may also be used and be of interest. These sheets can then be roughened using fine sandpaper of varying grit size. After cleaning, the contact angles may be measured as a function of grit size. Droplets may be obtained by condensation when holding the surface over a steaming vessel, or by using a pipette.

Using a camera or a webcam, together with a suitable background to the drop and the surface, you can measure the contact angle by a shadow technique. There are free software packages to analyze the camera images of the droplet on the surface [3].

It is also of interest to study contact angles as a function of drop size. Roughening methods other than sandpaper could be considered (e.g. etching), as well as various microscope techniques to measure surface roughness more accurately. Some old theoretical papers [4,5] on contact angles are still relevant and should be consulted.

REFERENCES

1. B.J. Ryan and K.M. Poduska, Roughness effects on contact angle measurements, *Am. J. Phys.* **76**, 1074 (2008).
2. G. McHale, Cassie and Wenzel: Were they really so wrong?, *Langmuir* **23**, 8200 (2007).
3. M. Brugnara, Contact angle, ImageJ plugin available online at <http://rsbweb.nih.gov/ij/plugins/contact-angle.html>.
4. R.N. Wenzel, Surface roughness and contact angle, *Ind. Eng. Chem.* **28**, 988 (1936).
5. J.A. Young and R.J. Phillips, An indirect method for the measurement of contact angles, *J. Chem. Educ.* **43**, 36 (1966).

Light scattering by aerosols, haze testing

The theory that applies here is that of Mie scattering. A related subject on differential light scattering in liquids was dealt with in Chapter 33. Initial experiments and probes could be very simple: construct an apparatus with a light source, a tube into which scattering particles in differing concentrations and dimensions may be introduced, and a photodiode as a light sensor, together with the appropriate electronics for amplification and recording the scattered light intensities. Part of the work is calibration of the light sensor and establishing its linearity. An important aspect is figuring out how to control the concentration of scatterers.

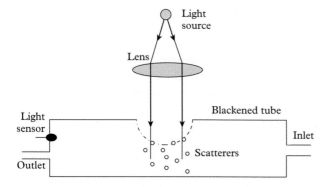

Figure A.9 *Experimental setup for aerosol scattering.*

The particles could be polystyrene microspheres, sand, or chalk dust. A possible initial simple arrangement is shown in Fig. A.9.

The project can be upgraded in sophistication and accuracy if appropriate optics equipment (spectrophotometers, integrating spheres, etc.) are available. For guidance consult the extensive literature on atmospheric applied optics, starting with Refs [1–5].

....

REFERENCES

1. H.R. Green and W.R. Lane, *Particulate Clouds: Dusts, Smokes and Mists*, 2nd ed. (van Nostrand, Princeton, NJ, 1964).
2. E. McCartney, *Optics of the Atmosphere* (Wiley, New York, 1976).
3. C.F. Bohren and D.R. Huffman, *Scattering of Light by Small Particles* (Wiley, New York, 1983).
4. R. Okoshi et al., Size and mass distributions of ground-level sub-micrometer biomass burning aerosol from small wildfires, *Atmos. Env.* **89**, 392 (2014).
5. D.B. Curtis et al., Simultaneous measurement of light-scattering properties and particle size distribution for aerosols: Application to ammonium sulfate and quartz aerosol particles, *Atmos. Env.* **41**, 4748 (2007).

A flag flapping in the wind

See Refs [1–4]. Cloth or polythene flags blown by wind have a flapping motion about both a vertical and a horizontal axis. A flag made out of packing foam will flap about one axis only, and this case is easier to study quantitatively. Construct an apparatus consisting of such a suspended flag and a well-directed variable-speed air flow. The latter could be a strong fan driven by a variable autotransformer, its air stream regulated by a honeycomb arrangement usually found in a wind tunnel, but which may be homemade. The wind speed could be measured by the deflection of a ping-pong ball suspended in the air stream.

Investigate the frequency and amplitude of the flag's motion as a function of wind speed. Use dimensional analysis to predict the frequency in terms of the important parameters of the problem, some of which may be varied in the experimentation: the flag areal density ρ, its length L, the air viscosity μ, the wind speed V, and some roughness factor C_R associated with the flag in the stream. Check whether some power law behavior holds for the speed dependence of the amplitude A: $A \approx V^n$.

Consult papers [1,2] on flexible plate behavior in an airstream for theoretical guidance.

REFERENCES

1. G. Houghton, The origin of lift forces in fluttering flight, *Bulletin of Math. Biol.* **28**, 487 (1966).
2. L. Tang and M.P. Paidoussis, Cantilevered flexible plates in axial flow, *J. of Sound and Vibration* **305**, 97 (2005).
4. A. Wood, *The Physics of Music* (Methuen, London, 1962), p. 115.
5. L. Prandtl, *Essentials of Fluid Dynamics* (Blackie and Sons, London, 1967).

Mechanical properties of paper

This is a vast subject, of interest to the paper production industry [1,2]. A start can be made by dropping steel balls of various diameters and mass from varying heights on circular samples of paper suitably held and clamped horizontally, to see how breakage depends on the kinetic energy and size of the balls. Some visual or optical criterion should be adopted which would reproducibly signal the beginning of fracture. For small radii there must be a "wrapping" effect on the ball at impact. Tensile and shear strengths may be investigated, as well as the influence of controlled amounts of moisture absorbed in the paper on the mechanical properties.

REFERENCES

1. H.H. Espy, The mechanism of wet-strength development in paper: A review, *TAPPI Journal* **78**, 90–99 (1995).
2. Institute of Paper Science and Technology, Georgia Institute of Technology, Atlanta, GA. Research reports available online at <http://www.ipst.gatech.edu>.

Gas discharge from pressurized tanks

This project monitors the tank pressure as a function time, $P(t)$, when air or carbon dioxide is discharged from a pressurized tank through a narrow tube whose length may be varied at will. The nature of the gas flow will change, since the Reynolds number Re will be a function both of the time elapsed and of the tube length. The basic equipment needed is a pressurized tank, a suitable valve, a pressure transducer controlled by a data acquisition system, and the tube through which the gas discharges. At different stages of the flow different theoretical flow models may be relevant. For $Re < 1000$, irrotational viscous flow (Poiseuille) dominates. At higher Re friction with the pipe wall becomes important, with a friction coefficient given by Moody charts [1,2]. Finally, for very short tubes they begin to act like an orifice. It is of interest to solve the differential equation for $P(t)$ in each of these cases [3] and to compare with experimental findings.

....

REFERENCES

1. Tools and basic information for design, engineering and construction of technical applications, available online at <http://www.engineeringtoolbox.com>.
2. Wikipedia, Moody chart, available online at <http://en.wikipedia.org/wiki/Moody_chart>.
3. Wayne State University, Mechanical engineering PhD preliminary qualifying examination – fluid mechanics, available online at <http://engineering.wayne.edu/me/pdfs/pqe_fluid_w02.pdf>.

Oscillating candle

A candle is pivoted horizontally about a needle passing through it near its center of mass. When the candle is lit at both ends, it will start oscillating. Investigate this thermo-mechanical system and determine the optimum choice of parameters for which its mechanical power will be a maximum. Reference [1] gives a useful theoretical model and its experimental verification. Several YouTube videos (search for "video demo burning candle at both ends") are available for demonstrating the phenomenon.

....

REFERENCE

1. S. Theodorakis and K. Paridi, Oscillation of a candle burning at both ends, *Am. J. Phys.* 77, 1049 (2009).

Properties of a flame

We suggest a series of possible explorations – see also Refs [1–3].

Place a candle between the oppositely charged poles of an electrostatic machine; the flame is attracted toward the negatively charged pole and repelled from the positively charged one.

One of the products of the burning of the hydrogen released by the vaporized fuel is water. You can collect it by passing a cold spoon over the top of your candle's flame, taking care to avoid accumulating soot. To measure the amount of water, put a funnel over the flame and collect the water in a cool flask by means of a rubber tube. You might want to measure how much water is produced by a given amount of depletion of the candle's wax.

Place the candle in strong sunlight and examine the shadow cast by the flame. The darkest portion of the shadow comes from the luminous region, which is the brightest area of the flame itself. The darkness of the shadow results from the fact that the collection of solid particles is densest there.

Show how combustible the vaporized hydrocarbons of a flame are. Carefully blow out the flame of a candle with a quick exhalation, but do not otherwise disturb the rising stream of white vapor that remains. Hold a lighted match above the wick and in the rising stream. If you do it soon enough, the flame will leap downward from the match to the wick and relight the candle. When you blow out the candle, some of the wax continues to be vaporized by the remaining hot gases and the wick. It is this rising stream of vapor that is combustible.

By making or buying a series of candles that differ only in the diameter of their wick you can demonstrate the possible types of candle flame, from the kind that barely keeps burning because of poor fuel flow to the kind that smokes abundantly because there is too much fuel flow. Examine the characteristics of the flames. The dimmest possible flame is entirely a reaction zone and thus has only blue light. In a larger flame the luminous carbon region begins at the top of the reaction zone or just inside it and grows upward to split the reaction zone apart. In a higher flame the blue reaction zone is around the base, and the top of the luminous carbon zone is exposed directly to the outside air. Eventually the flame rises so high that the inner core of the luminous carbon zone cools; the carbon particles are no longer consumed by the time they reach the top of the flame, and so they are released as soot.

Check the spectrum of a candle's emission with an inexpensive diffraction grating and with a glass prism. The luminous carbon zone dominates the light emission, and you will see a complete spectrum from red to deep blue, with no hint of any molecular emissions. To see the C and CH emissions use a spectrometer of a sort commonly available in school labs. The spectrometer has a movable telescope with which to examine the dispersed light. Light from the candle enters a narrow slit and is directed by a lens onto either a diffraction grating or a prism, which will deflect and disperse the light to show individual colors.

When light from the luminous carbon zone enters the slit, a complete visible spectrum will be seen. When only the light from the blue reaction zone on one side of the flame enters the slit, the full spectrum from the incandescent particles is not as bright, and you will distinguish several molecular emissions. Once you note the angular positions of these emissions you can calibrate the spectrum of colors in terms of wavelengths by removing the candle, replacing it with a sodium lamp, and noting the angular positions of the sodium emissions. The wavelengths of sodium emissions are listed in reference books; using these, you can then identify the molecular emissions as being from the C and CH molecules by calculating the wavelengths associated with the angular positions you have recorded and then referring to previous work listing the wavelengths of the emissions from those molecules.

REFERENCES

1. H.A. Wilson, Electrical conductivity of flames, *Rev. Mod. Phys.* **3**, 156–89 (1931).
2. M. Faraday, *The Chemical History of the Candle* (1860). Available online at <http://www.fordham.edu/halsall/mod/1860Faraday-candle.asp>.
3. J. Walker, The physics and chemistry underlying the infinite charm of a candle flame, *Sci. Am.* **238**, 154 (1978).

Cooling water

Two identical open glasses, filled with hot and warm water, respectively, begin to cool under normal room conditions. Is it possible that the glass filled with hot water will ever reach a lower temperature than the glass filled with warm water? Design an experiment to investigate this and explain the result. Can hot water freeze faster than cold water? In constructing a suitable apparatus, consider the possible influences of the environment, supercooling, convection currents, dissolved gases, and evaporation on the cooling process. See also Refs [1–3].

REFERENCES

1. E. Mpemba and D. Osborne, Cool?, *Phys. Ed.* **4**, 172 (1969).
2. I. Firth, Cooler, *Phys. Ed.* **6**, 32 (1979).
3. D. Auerbach, Supercooling and the Mpemba effect: When hot water freezes quicker than cold, *Am. J. Phys.* **63**, 882 (1995).

Figure A.10

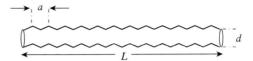

Figure A.11

Corrugated plastic pipes: the whirly

These toys [1] are like vacuum cleaner pipes, corrugated along their lengths. When rotated they emit sounds of varying pitch, depending on the speed of rotation. The basic parameters are the pitch size a, the diameter d, and length L of the tube, and the rotational speed ω – see Fig. A.10.

Of interest is the fact [2] that the sounds emitted arise at only certain discrete frequencies f_n, each of which occurs for a restricted range of angular speeds $\Delta\omega_n$. A way must be found experimentally to rotate the tube at precisely controlled speeds. One way is to insert the tube in an open container resembling a diving bell, as shown in Fig. A.11. This container is then pushed down and lowered at controlled speeds (measured with an overhead sonic ranger) into a large tub full of water, while the sound emitted by the tube is registered by a microphone connected to a data acquisition system. The data are then analyzed by an FFT program in order to determine accurately the frequency.

Alternatively, the tube may be fixed to a rotating bicycle wheel which is rotated by an electric motor, arrangement being made to prevent entanglement of the associated sensor connections, as in Ref. [3]. Varying the parameters experimentally and modeling the behavior of the resonant air column in the tube [3], paying particular attention to the repeated collision of the air with the corrugations, makes for an interesting project.

....

REFERENCES

1. P. Doherty, Whirled music, 2001, available online at <http://isaac.exploratorium.edu/~pauld/activities/AAAS/aaas2001.html>.
2. F.S. Crawford, Singing corrugated pipes, *Am. J. Phys.* **42**, 278 (1974).
3. M.P. Silverman and G.M. Cushman, Voice of the dragon: The rotating corrugated resonator, *Eur. J. Phys.* **10**, 298 (1989).

Sound from a boiling kettle

See Refs [1–4]. With no more equipment than some cylindrical beakers, kettle, microphone, thermocouple, and spectrum analyzer, one can analyze the sound spectrum of the noise emitted by water in a heated vessel as it approaches boiling and thereby disentangle the various mechanisms which contribute to the noise spectrum. The sound will be a function of the shape, material, and size of the vessel, the initial amount of water in it, as well as the temperature as boiling is approached. Appearance and disappearance of modes in the spectrum, and their possible shift as the amount of water in the vessel is varied, will indicate whether one is dealing with resonance modes of the air cavity above the water, flexural modes of the walls of the vessel, nucleate boiling with small bubbles of dissolved air in the water rising to the top, vapor bubbles forming and explosively collapsing at the bottom of the heated vessel and thus exciting acoustic modes, or, in the final stages, vapor bubbles growing, coalescing, and forming jets and columns stretching from the heated bottom to the surface. Full boiling is characterized by the formation of large vapor bubbles throughout the bulk of the liquid. Water may be pre-boiled to eliminate dissolved air. Differences between tap and distilled water may be investigated. The effect of a layer on oil on the surface may be checked.

....

REFERENCES

1. S. Aljishi and J. Tatarkiewicz, Why does heating water in a kettle produce sound?, *Am. J. Phys.* **59**, 628 (1991).
2. W. Bragg, *The World of Sound* (Dover, New York, 1968).
3. M. Kummel, The sound of boiling water, 2001, online at <http://www.rain.org/~mkummel/stumpers/28sep01a.html>.
4. M. Minnaert, On musical air bubbles and the sound of running water, *Phil. Mag.* **16**, 235 (1933).

Appendix B
Facilities, materials, devices, and instruments

This appendix describes what is needed in a project lab generally; each chapter of the book lists what is needed for individual projects. What is available has a role in the project from start to finish. For schools intending to set up a project lab which will be a permanent part of the curriculum for the whole class, it is useful to give an overview of the arrangements for such an enterprise, ranging widely from, say, plumbing parts and tools to signal processing and spectrum analysis. Most of what follows applies to this case and much will be familiar to many readers. As for what is *not* familiar, we believe that for both the guide and researcher there is a difference between *no* experience or knowledge and even a very little. If you've never seen, heard, felt, or read about something, you can't look it up. For example, consider the difference between never having held a claw hammer, and having held it and driven ten nails and pulled out five. You could then look up hammer and then nails on the web or eBay and see what there is. The same is true with unfamiliar items that are listed throughout this appendix.

Facilities

Space

Set aside a separate space, so that projects in progress will not be touched or disturbed, except for occasional borrowing of equipment. Some darkroom facilities and acoustically insulated space are desirable. Each pair of investigators needs a sturdy table with drawers, storage space, and half a dozen electric outlets. The table drawers might contain, say, eight basic tools: two pliers, two screwdrivers, one wire cutter, one tweezers, one wire stripper, and one soldering iron. This is complemented by an extensive tool pool, contents listed below, kept and provided on demand by the technician. The technician must have a personal area, preferably adjacent to the common workshop, with power tools, use of which he/she guides.

Tools and workshop

We list a collection of useful tools that we have found invaluable. Many were hung on perforated wallboard, painted with tool outlines, to ensure orderly replacement after use.

- Small crosscut handsaw, hacksaw (with fine- and coarse-toothed blades), coping saw.
- Block plane.
- Files: flat, triangular, round, half-round, fine, coarse, mill, bastard, rasp.
- Screwdriver sets (flat and Phillips), including miniature and jeweler's.
- Various pliers.
- Wrenches (both hand and ratchet driven) of many sizes and types: crescent, open-ended, box, small pipe, socket.

In addition, the workshop would benefit greatly from the following:

- A workbench with a vise and fixed power tools, simple and small, including: drill press, band saw, grinder (with stone and wire scratch wheels), sander (belt and disk).
- A small lathe (possibly combined with a simple milling attachment).
- A small, accurate vertical drill on a tabletop stand.
- An electric table saw (used with supervised access).
- Twist drills, taps and dies, hole saws, metal and plastic pipe and tubing cutters, flaring tools, and reamers, all to be kept in drawers.

Skills

The beginnings of skills can be obtained from reading, watching others, being guided, and simply fumbling around; no one becomes expert, but a broad, extendable view is acquired. Here we describe one skill which is invaluable, soldering.

Soldering

The need for soldering wires and electronic parts will be ever present. Solder works somewhat like glue on wood – glue gets into the fibers of the two objects to be joined and bonds them when it hardens. Hot, molten tin–antimony alloy solder alloys itself with the surfaces to be bonded, making a strong joint when it cools and solidifies. Soldering is a quick and easy way to repair or put together all sorts of things (made of iron, tin, copper, or brass, but not aluminum), repairing leaky pots, making tin flowers, connecting electrical wires, and so on. Soldering flux, applied to the things to be soldered, is used to clean the surfaces chemically. Sometimes flux is in the hollow core of soldering wire and sometimes applied separately from a small container.

A few minutes after being plugged into a wall outlet your soldering iron will heat up to soldering temperature. When you touch solder that has been dipped in flux to the hot tip, the solder should wet the tip and flow around it. If you keep pushing solder onto the tip, a molten blob will finally drop off and go *splat*. Your solder's melting temperature is around 200°C. Your iron will heat to maybe 100° higher.

Molten solder must *wet* the metals to be joined for proper bonding to occur. To get the idea, look at water drops on a greasy surface that is not wetted and on a clean surface that is. Molten solder won't wet dirty, corroded, or oxidized surfaces. To dissolve that oxide and to prevent the formation of more, you use flux that smokes when hot; the smoke is not harmful in reasonable amounts. Tests also show that the hazard from solder vapor from soldering on this scale is negligible.

A new soldering iron should be *tinned* – covered with a shiny film of solder – as it heats up for the first time, to prevent corrosion. Cover the entire tip surface with fresh molten solder and then wipe it quickly with a damp cloth or sponge. The solder should flow onto the tip readily. If the tip gets pitted and corroded after some use, it will appear black and soldering will become difficult: there will be no liquid surface to conduct heat efficiently from the iron to what you are soldering. To improve matters, cool the iron and try filing or sanding the tip clean, then reheat and tin with fresh solder before the tip re-oxidizes. This may fail on plated steel tips, but tips can easily be replaced.

Soldering depends on heating both parts to be joined, using a drop of molten solder between the iron and the work to conduct heat from the iron to the parts. Try soldering single and multiple strands to each other and to a thin sheet of metal cut out with tin snips. Most of the materials will solder readily. If they appear dirty, clean them with sandpaper, or use a screwdriver or a pocketknife as a scraper. Then re-tin.

The finished work should be smooth and hard to pull apart. As you gain in skill, you will probably find yourself using less solder. You control the amount by how much solder you push onto the joint.

Notice that things get hot, including *all* of the metal parts of the iron – use pliers or heat-insulating stuff to prevent burnt fingers, and remember that your solder melts at around 200°C; the iron can also melt plastic tool boxes, burn you, your desk, paper, cloth, and so on. Be cautious without being timid. Unplug the iron when you're done. Use the stand. Keep your work area clear. Know where your iron is! And make sure that it's off when not in use.

A Google search for "basic soldering techniques" will enable you to watch a number of demonstrations and tutorial explanations of the right soldering methods on YouTube.

Materials and storage

After a few years of operating a project lab a great deal of material will accumulate, reducing trips to stockrooms and stores as well as ordering. This material needs to be stored in an orderly way. Overhead shelves and cupboards with glass doors, large cupboards, tables with boxes to contain miscellaneous items from lengths of stock to recyclable parts from the past, can all be installed. It is important to be aware of what there is in the growing collection. So keep material, informally inventoried or listed, in labeled boxes; use cabinets for instruments and open boxes for "junk" and salvage. This enables researchers to cruise around for inspiration in designing and building their apparatus. Many things in the following list in our schools were not acquired systematically – they simply piled up and some might well be considered junk. The items in italics are particularly valuable in the rapid improvisation and assembly of apparatus.

- Metal: sheet, rod, tube, pipe, angle (Dexion, structural), different materials (iron, steel, stainless, copper, brass, lead), ball bearings, and steel balls.
- Many kinds of C or G *clamps* and *corner irons* in a wide range of sizes.
- Wood: boards, studs, plywood, Masonite, dowels, sheetrock, appropriate glues.
- Glass: window or sheet glass, mirrors, Pyrex tubes (thin wall and capillary), rods.
- Plastic: sheet, rod, tube, pipe, different materials (Lucite, Teflon, Bakelite) plus cements.
- Rubber: sheet, tubing, balls, gaskets.
- Flexible tubing: vacuum and garden hose, clear plastic, copper.
- Fittings: pipe, tube, plumbing adaptors, couplings, tees, ells, clamps.
- Clamps: stainless steel *screw hose clamps* invaluable in holding apparatus together.
- Hardware: Many kinds of *corner irons* in a wide range of sizes; hinges, nails, tacks, screws and bolts with nuts (also wing and elastic stop nuts), washers (flat, fender, lock) in a wide range of types and sizes in labeled drawers in cabinets, also "hell boxes" of miscellany, also "chemical hardware and glassware": ring stands, flask supports, right-angle clamps, stop cocks, flasks, test tubes.
- *Wire coat hangers* are a great source of iron wire to use in all kinds of apparatus.
- Wire: lamp (zip) cord, various gauge insulated and bare, stranded and solid copper wire, resistance wire, single- and multiple-conductor shielded cable, coaxial cable.

- Permanent magnets: all sorts and sizes – bar, horseshoe, round, disk, flexible, also from junked microwave ovens.
- Electric: household fixtures of all kinds.
- Fittings: for all wire and cable types, wiring nuts, terminal strips, sockets, plugs, clamps.
- Electronic: perfbord, stripboard and protoboard, transistors and ICs with sockets, resistors (fixed and variable), capacitors, inductors in labeled drawers in cabinets, also "hell boxes" of miscellany.

Devices

This section lists and briefly describes much of what is useful in a project lab that is neither a tool, material, nor an instrument. Again, store such items so that they can be readily located.

Transformers

In recent decades discarded digital devices provide a plentiful source of < 25 W wall transformers at various AC voltages. Note that a soldering gun can deliver over 100 A at fractions of a volt. A variable autotransformer (such as a Variac) in the 0–240 V range is of great utility. Transformers from junked radios and TVs and for neon signs and oil burner ignition, as well as auto ignition coils, can deliver voltages from a few hundred to over 15,000 V. Note that 10 to 15 kV produces a spark about a centimeter long between wires. Audio transformers from radios of all sizes are useful for handling low frequency signals.

Power supplies

Again, discarded digital devices have power supplies with unregulated 5 to 25 V DC output at less than 25 W. The power supplies driven by the salvaged transformers previously mentioned have many applications; suitable high voltage or current rectifiers and filter capacitors can be added. Researchers may also build their own LVPS (low voltage power supply, 1 to 12 V adjustable, 1 A regulated) and HVPS supplies (high voltage power supply, 100 to 1000 V adjustable, internal resistance 1.5 MΩ) from the circuit diagrams at the end of this appendix. Such an exercise is highly recommended.

Amplifiers

Audio power amplifier from public address systems, or homemade from transistors and/or op-amps. There are easily located circuit diagrams in the references and on the web for a large variety of amplifiers. They can often be built by researchers themselves for a particular application, at specified amplification and noise levels. The circuit diagram of one we have frequently used is at the end of the appendix. Again, this exercise is recommended, where appropriate.

Actuators

To exert forces on a system from an electric or a pneumatic source: collect AC and DC motors, relays, solenoids, armatures, and wound coils of all sizes. Coils of wire in stores wound on plastic spools, when both ends are accessible, make ready-made solenoids that can act on iron slugs. Loudspeakers of all sizes can be used as drivers if only small forces are needed. Compressed air from a movable nozzle directed into a hole can, in effect, be a switch or an amplifier of force. Plunger-like items such as bicycle pumps or door closers can exert large forces when connected to compressed air.

DC motors are so useful as actuators, for turning things, and as sensors, measuring angular velocity, that the following mini-project can be worthwhile: use a short piece of rubber insulation from a 5 kV test lead as a flexible coupling between 2 small DC motors mounted with clamps on a piece of wood. Drive one motor from an adjustable LVPS and light a lamp or use a resistor as the load on the second motor, now functioning as a generator. Measure everything, compute efficiency, and so forth.

Sensors (commercial)

These are often central to projects. There are many computer-operated data acquisition systems on the market in the US, UK, Germany, and so on. Among the widespread brands are (in alphabetical order), Fourier, Leybold, Pasco, Phywe and Vernier. Most can provide a whole range of sensors: motion (both linear and rotational), force probes, voltmeters, magnetic field, temperature, pressure, pH, light intensity, sound field (microphone), and so on. All these operate through a computer interface. Their capabilities include sampling rates from 20 kHz to 100 kHz or beyond, spectral analysis (FFT), simultaneous multiple sensor applications, signal generators, oscilloscopes, power sources, all rolled into a single package, together with convenient and user-friendly data analysis programs. When choosing a system, these are the parameters that should be compared and weighed carefully, and of course the price, some very high. If the users do not know or care how these sensors work, there is the additional "price" of being ignorant of the principles of operation.

Sensors (home-made)

The following can easily be built by the researchers themselves, saving money and enhancing understanding: a small search coil for magnetic field mapping; a light meter from a photodiode or phototransistor (available for detecting radiation having a wide range of wavelengths); a position sensor; a sound sensor.

Here are some simple examples of homemade sensors. A position sensor (not low friction) can be built from a ten-turn potentiometer with a reasonably large shaft, together with a small metal pulley or a thread spool and thread which is wrapped around the pulley to avoid slip. Attach the thread to the moving object, such as a falling object. Connect a 1.5 V battery to the two outer terminals of the pot and calibrate by measuring the voltage between the slider and one outer terminal as a function of angle of rotation (or position if pulley and thread are attached).

A three-pole armature, permanent-magnet motor which, with a pulley and thread and acting as a generator or tachometer, becomes a velocity sensor. It can be calibrated by driving the tachometer with another motor and observing the average DC output versus the frequency of the ripple associated with commutator action, one sixth of this frequency being the shaft speed in revolutions per second.

A microphone or small loudspeaker can be used to display the sounds of bouncing balls for analysis in terms of the coefficient of restitution.

Small AC two-phase capacitor induction motors can be similarly used; excite one phase from the mains and observe output from the other. Loudspeakers can be used as sensors of vibration or as vibrators in mechanics experiments. Putting a mass on the voice coil and measuring its capacitance to the frame makes a crude displacement or acceleration meter. Small, readily available, condenser microphones are versatile sound sensors. These sensors will have voltage outputs that can then be fed to a data acquisition module with suitable converters. An LC oscillator in which a coil spring is used as the inductor will change frequency as forces deflect the spring, the changes detectable by an FM discriminator. The same arrangement can be used to detect changes in capacitance when capacitor plates move, or if a dielectric comes between them due to forces, fluids, and so on.

The use of simple sensors with an oscilloscope and some accessories was part of a student seminar and would be appropriate for a project lab. Shortened instructions on two lab exercises are given at the end of this appendix, one on using simple sensors, another on the failure of electrical components due to overload, or blowout.

Noise and band pass filters

When signal output from any device is to be observed, measured, or recorded, it may be desirable to use simple filters to reduce noise, or to select particular frequency bands. These include high- and low-pass single and multiple RC filters and twin tee notches (to suppress power-line hum). RC circuits can do an approximate integration (averaging) or differentiation of transient signals or pulses. A notch in the negative feedback loop of an op-amp makes a useful frequency-selective amplifier and, if tunable with a dual pot, a primitive wave (spectrum) analyzer.

Instruments – mechanical

The following are the basic mechanical tools: tape measures, meter sticks (incidentally, we have compared wooden and metal meter sticks and found variations of a few millimeters between them, as might be expected), rulers, calipers (divider, inside, and outside), vernier (understand how the vernier works) and dial calipers, depth gauges, micrometers, stopwatches, clocks, balances (spring and weight). These things may sit around for a long time, so it's best if they don't have a digital display with batteries to run down. Spring, pan, and weight balances, from the delicate chemical types, through home floor types, to those used in factories and warehouses are all useful when needed. A surveyor's transit (theodolite) as used to measure angles and elevation, generally outdoors, was once part of a project.

Instruments – electrical

Many projects involve electrical devices to provide power, often from signal generators, to drive the project system or apparatus to produce forces, mechanical vibrations, sounds, electricity, light. Another set of devices is needed for detecting response and acquiring, recording, and analyzing data. It is a rare project that pushes current instrumentation limits; the majority deals with times longer than a tenth of a microsecond, and frequencies less than 10 MHz. In what follows we discuss the basic instruments that should be available for projects involving electrical devices: multimeters, oscilloscopes, and signal generators.

Multimeters

Many projects use several multimeters, most likely low-cost digital ones (DMM). Having them on hand is helpful both for both diagnosis and for measurement, since so many projects involve electrical devices.

Many low-cost DMMs are on the market, all operating on internal batteries. A typical one has a three- or four-digit digital display (depending on the range measured) and a five-position selector switch for on/off, DC V, AC V, diode check, KΩ, and continuity. Red and black test leads with probes are permanently attached. There are warnings not to exceed 400 V AC or DC. The auto ranging system gives readings from hundreds of volts to millivolts and from 1 MΩ to tenths of an ohm in well under a second. There is no ohms-zero adjust and many meters indicate a few ohms with the test leads shorted.

The ease and speed with which readings can be made is one great advantage of the DMM. Also, below DC volt readings of a few tenths of a volt the input resistance becomes extremely high, so that a charged 0.1 μF capacitor remains charged for minutes. This feature can be used to average or integrate a DC signal and for timing events by using a switch to connect a battery and series resistor to the capacitor.

The project lab should also have some high-grade DMMs having dB, transistor gain and diode check, capacitance, and frequency measuring capabilities. They are helpful when precision, high resolution, versatility, or speed of reading is needed, and for calibration of other meters. Inter-comparison can weed out obviously faulty meters, but the question of calibration could be a project in itself. Consider a typical 10 times more costly DMM. It also has a three- or four-digit digital display and below that a horizontal 40-division display showing magnitude and trends in a way that assists in making adjustments for maxima or minima. Besides an on/off switch there is a multiple-position selector switch for temperature (used with supplied probe), continuity, diode check, ohms, capacitance (in nF), DC V (1000 max), AC V (750 max), DC current (400 mA max, up to 10 A through a separate jack), AC current (400 mA max), frequency (in kHz), and transistor gain (h_{fe}). There are also push buttons to choose dB, ranging (automatic, or manual up or down) and hold (to keep a reading), and Fahrenheit/Celsius. There are banana jacks for the plugs of red and black test leads with probes, as well as small sockets for transistors, diodes, and capacitors. It uses internal dry cells or a wall plug-in supply. Although in some DMMs the internal cyclic comparison processes result in low-level RF signals around the meter, this is not usually a problem.

Very few projects are likely to need precisely calibrated meters (better than 1%), but if so, Josephson voltage standards and Weston cells, if available, can be used. An assortment of precision resistors may be helpful for some experiments.

If analog multimeters (AMM) are still around, use them. Their advantage is that, unlike the DMM, their working can be understood quite thoroughly by the researchers, including the diagram showing the multi-position double-pole range selector switch with shunt and multiplier resistors and rectifier.

We describe here various possible activities with an AMM, whose sum will certainly add up to a worthwhile project.

- Swing the meter horizontally around a vertical axis and compare the damped needle motion when the switch is in the "off" setting, less damped on any other setting.
- A typical meter has DC and AC voltage ranges with internal resistance of 20 KΩ/V and 10 KΩ/V, respectively. Remove the 1.5 V battery and bypass the battery holder terminals with a clip lead. Use another meter to read resistances on various ranges, resistance, DC and AC voltages. Note the readings of both meters; do they make sense?
- Understand the loading effect of the voltmeter depending on circuit impedance, and the effect of lead and shunt resistance on current measurements.
- Measure currents and voltages with two meters. Do they agree?

- Improvise other ranges: A series 15 kΩ resistor with the meter on 50 µA/ 250 mV gives a 1 V full-scale range. The 5 V range with a 9 V battery gives a resistance range on which 100 kΩ reads at "7" instead of "100" on the resistance scale. Some calculation is needed to determine the resistance. A capacitative divider can be used on the AC ranges. Use a capacitor whose reactance at the frequency involved is small compared with the meter resistance to bypass the meter. Connect the high AC voltage to be measured through an appropriately small capacitor with a suitable voltage rating. For the latter here is a homemade version: two metal plates separated by air, or material with known dielectric constant, a plastic or mica. For measuring currents of a few amperes a length of resistance wire (example: diameter 1.3 mm, Ni, Cr, Al, Cu alloy, 1 Ω/m) can be used as a shunt. Be sure to separate current and voltage leads.

- Show that a plot of numbers on the ohms scale (resistance R) versus those on a voltage scale (in effect, current i), excluding $R = 0$ and $R = \infty$, is an inverse exponential:

$$i = c_1 \exp(-c_2 R).$$

- Present and analyze data on the following: moving coil wire size, number of turns, resistance; magnetic field; spiral springs; coil bearings; needle deflection and response dynamics; range switch and circuitry, multipliers, shunts, rectifier; precision, comparison with quality meters, standards; effect of meter internal resistance on measurements; effects of test leads.

Miscellaneous meters

The following may have roles in some projects: panel-mount meters with full-scale DC ranges from 200 µA to 20 A (a shunt may be in this one) and 5 to 200 V, AC ranges 0.5 A to 20 A and 5 to 200 V; clamp-on ammeters with ranges 10 to 100 A; vibrating-reed frequency meters, 55 to 65 Hz; an analog household wattmeter or a digital miniature wattmeter.

Connectors

There are many kinds of cable and wire connectors. For banana plugs, often in red (high) and black (low or ground) with an outer plastic screw for wires, two of these mounted on plastic and ¾ in apart form a GR plug (found on many instruments originally made by General Radio, a now defunct company). There are UHF, type N, and BNC plugs and the current TV form, all working with cables that should be properly terminated −50, 75, 300 Ω being common. There are different size phone plugs, mono and stereo, and the ubiquitous RCA connector. Different projects will have different needs, but copious supplies of cables, connectors, and adaptors should be available, along with simple specialized tools for attaching connectors to cables.

Oscilloscopes

These are very helpful both for diagnosis and for measurement for any project where electrical signals and power are involved. The habit of using them to look at everything to see what sorts of time variations are present is important. Does the power supply have ripple? Do pulsed circuits react on battery supplies? Is there hum or extraneous noise on the signal from ground loops, bad shielding, fluorescent lamps, and so on? Is there limiting due to overload?

Oscilloscopes, both analog (AO) and digital (DO), are available with a wide range of bandwidth, sensitivity, number of controls, ease of use, size, and cost. A high-grade oscilloscope is helpful when its extra performance or features are needed. There are bench-top and hand-held oscilloscopes, and plug-in attachments and associated software to be used with computers. A major manufacturer (Tektronix) has an online article, "XYZ of Oscilloscopes," that is very informative.

A typical low-end AO has 20 MHz bandwidth, horizontal (X) and vertical (Y) sensitivity to 1 mV per division, sweep to 0.2 µs per division, two BNC inputs plus one for external trigger (with GR adapters). Here is a description of a brief exercise for researchers who have never used an oscilloscope, given at the start of the project lab, in order to familiarize them with this most useful display device.

- Use wire cutters, wire strippers, and soldering iron to make a clip lead from a length of wire, alligator clips, and insulating sleeves.
- All controls on the oscilloscope are off or turned fully counterclockwise. Find, center, and focus the spot and move it in horizontal (X) and vertical (Y) directions.
- Connect the 1.5 V battery to the inputs and observe the deflection of the spot in the X and Y directions with different settings of the range switches such as 1, 2, 5, 10 V/div.
- One researcher moves the spot in the X direction while the other connects and disconnects the battery; subsequently, use the internal sweep to display the deflection in time.
- Wire a double-pole double-throw switch as a reversing switch for the battery connected to the Y input, and make a square wave by rapid switching, observed with internally triggered horizontal sweep; compete to see who switches faster (about 5 Hz is usual).
- Look at the output of the oscillator/signal generator (20 Hz to 200 kHz in four ranges) with square and sine wave output with different sweeps, and become familiar with connecting high and low (ground) terminals together.

- Connect the low voltage plug-in wall output to the X input and, with Lissajous figures, check the oscillator frequency calibration at 20, 30, 60, 90, 120, and 180 Hz against the 60 Hz AC voltage from the transformer (make appropriate choices for 50 Hz). Note that the number of loops along the edges of a stationary pattern gives the frequency ratio.

While doing this exercise, ideas of how to observe signals and how to connect up circuits with agility and accuracy begin to grow. Self-confidence develops with mastery of the instrument (fortunately indestructible, except for burning the screen) with about 20 controls, simply by studying the labels, thinking, observing, deducing. Of course, one can always read the instruction manual.

Digital oscilloscopes have many more functions, they are compact and light, but how they work is far more obscure than is the case with analog instruments. They are available in an extremely wide range of performance and cost, and project labs should have some of all kinds, perhaps outdated and handed down from research groups. A typical medium cost and size has memory, FFT (fast Fourier transform) for spectrum analysis, 60 MHz bandwidth, X and Y sensitivity to 1 mV per division, sweep to 5 ns per division, two BNC inputs plus one for external trigger. The accompanying user manual, along with experimenting with the numerous combinations of settings and functions, makes learning its use fairly straightforward. There is an extension module that enables connection to a computer or printer and a programmer's manual that describes a large group of commands and queries for controlling the oscilloscope through an appropriate interface. Few projects are likely to need or use much of this array of features, but the ability to store transients, measure time and amplitude, add and subtract, and record signals is valuable. Hand-held digital oscilloscopes, available with a wide range of features and cost, are practical complements to the ubiquitous multimeter, reserving the bench types for more demanding applications.

Signal generators

These are useful in driving and testing all sorts of apparatus. Low- to medium-cost instruments are satisfactory. Consider two typical analog signal generators, one low and one high in cost. One, suitable for mechanical, audio, and ultrasonic work, has four ranges from 10 Hz to 100 kHz with an output control and output range switch providing 5 mV to 5 V into 600 Ω, and two terminals, one grounded. The other, suitable for low-level RF work, has five ranges from 100 kHz to 100 MHz, with continuously variable output in two ranges: low, up to 0.5 mV, and high, to 0.5 V into 50 Ω. Consider also a low-cost audio signal generator that uses direct digital synthesis. It has two ranges, obtained either with a tuning knob or by pressing buttons: 0.1 to 599.9 kHz in 0.1 Hz steps, and 1 to 59999 kHz in 1 Hz steps. Output is either sine or square wave up to 1 V peak to peak. Any of these can drive some devices directly or through suitable amplifiers: stereo or public address types for audio, and likely homemade for RF. Finally, it's not hard to make homemade oscillators for special purposes; consult books of collected device diagrams. Some data acquisition systems will possess signal generator capabilities.

Instruments – optics

We list here some items that are necessary for optical projects.

Room

A dark room, with stable, sturdy, vibration-free table surfaces made from large cement or marble slabs resting on inflated tire inner tubes. A professional floating optical table is beyond the budget of most labs.

Light sources

These range from hot objects emitting IR to small bulbs to arcs.

- Low and high intensity LEDs
- Lasers, easily mounted and adjustable (also laser pointers)
- Optical bench incandescent lamps
- Fluorescent, UV, and Hg lamps
- High intensity lamps that approximate a point source
- LED flashlights
- Arc lamps (sometimes homemade using carbon dry cell electrodes) with R or L ballast.

Instruments

There are just a few that we have used in our labs:

- Optical benches and component holders
- Student prism spectrometer
- Photometer
- Black body chamber.

Components

Here the list is extensive, and the particular project will dictate the choice.

- Lenses, mirrors, collimators, apertures
- Filters, rotatable polarizers, front-surface mirrors

- Monochromators, prisms, gratings
- Spatial filters, precision XY tools
- Photodetectors, photodiodes to photomultipliers (cooled by dry ice if needed)
- Light power standards.

Instruments – heat

Most work with heat involves sources and methods of controlling and measuring heat flow and temperature. Here are listed some useful items:

- Thermal insulation of all kinds
- Vacuum flasks and Dewars
- Heat sources: IR, electrical (resistance wire, nichrome, or high-wattage resistors), combustion, usually gas
- Thermometers of many kinds and ranges, from household to calibrated Hg
- Thermocouples, iron (or copper) versus constantan, reference junction temperature, voltage measurement
- Thermistors of many sizes and ranges, positive and negative temperature coefficient
- Black body: paint, bundled sewing needles.

Activities

Here we describe briefly the two exercises mentioned above, namely with sensors and on blowout.

Sensors and an oscilloscope

Researchers met weekly to do experiments with various simple sensors, calibrating them with the oscilloscope and auxiliary apparatus. The principal sensors were:

A 50 kΩ ten-turn potentiometer that, with a pulley and thread, can act as a position sensor. It was calibrated by connecting a 9 V battery and observing slider potential as a function of rotation.

A three-pole-armature, permanent-magnet motor that, with a pulley and thread, and acting as a generator or tachometer, becomes a velocity sensor. It was calibrated by driving the tachometer with another motor and observing the average DC output versus the frequency of the ripple associated with commutator action, one-sixth of this frequency being the shaft speed in revolutions per second.

A small condenser microphone was used to record the sounds of bouncing balls for analysis in terms of the coefficient of restitution.

The pot and the tachometer were used separately and together to observe the motion of falling weights (both in free fall and on inclined planes) and of pendulums. Single-sweep curves on the analog oscilloscope were photographed to determine position and velocity versus time, or position versus velocity and thus acceleration, period, and the effects of friction.

Loudspeakers were used to measure impulse to stop falling weights, forces on a mass acting as a pendulum bob, and also to impart momentum to a mass and hence project it vertically.

An experiment to determine the acceleration due gravity g by using a speaker to accelerate a mass vertically is feasible.

Blowout

At some point, everyone blows out fuses and incandescent lamps. For example, it is common to blow out the fuse that protects an analog multimeter on its 250 mA range. With this in mind, we developed a short exercise in blowing out a fuse, a resistor, and a capacitor, using adjustable low-power voltage supplies (LVPS) for the fuse and to observe the heating of a resistor at its rated power before burning it out (as well as a 2 μF, 25 V electrolytic capacitor) with a 24 V, 2 A wall transformer. What happens?

- When 3.7 V from the LVPS is applied to a 27 Ω, 0.5 W resistor, the current is ~140 mA and the power is ~0.51 W; it gets quite warm. Current and voltage are measured using two separate AMMs.
- The LVPS, at a low setting (~1.5 V), is connected to a 0.25 A fuse in series with a 9.1 Ω 0.5 W resistor; raise the voltage until the fuse burns out, noting the maximal voltage across the resistor.
- Connect a 9.1 Ω, 0.5 W resistor (or 2.7 Ω, 0.5 W) to the 24 V transformer, so power is now over 60 W. Record the details of resistor destruction. Do the same with the capacitor. Ideally, the capacitor should be inside some safety enclosure (such as an upside-down glass jar). Wear safety glasses and watch out for hot vapors and heat.

Some useful circuit diagrams for homemade devices

Homemade power supplies

Building their own variable power supply is a valuable exercise for researchers embarking on any project, involving as it does some basic electronics, assembly skills, tinkering, soldering, understanding the principles making them work, and of course savings in comparison with commercial products. These devices can be put together either by plugging parts into a protoboard with no soldering, allowing salvaging and reusing components; or on perfboard: parts on one side and soldered connections on the other. Some of the components specified may no longer be available, but up-to-date equivalents can be found with a quick web search.

Low-voltage power supply (LVPS)

The power supply described here will provide stable DC voltages, variable from about 1 to 12 V. Components needed:

- Wall transformer, 220/120 V to 12 V AC/1 A
- Voltage regulator LM317T, with a heat sink
- Full-wave bridge rectifier, BR86D, with four leads
- Resistors: 240 Ω, 5 KΩ pot
- Capacitors: electrolytic 1 μF/50 V, 1000 μF/25 V
- Circuit board size UBS-100, on which the components are assembled
- Connecting wires.

The components are connected according to the circuit diagram in Fig. B.1.

The working principle of the LVPS is shown in the block diagram in Fig. B.2.

The wall transformer reduces the 220V AC (120 V in the US) from the line to a safe 12 V with only moderate loss of power due to heating of the transformer.

The full-wave bridge rectifier has four half-wave rectifiers, each of which allows current to flow in one direction. The 1000 μF capacitor then smooths out the ripple in the rectifier output. For a tutorial on the bridge rectifier, consult <http://hyperphysics.phy-astr.gsu.edu/hbase/electronic/rectbr.html>.

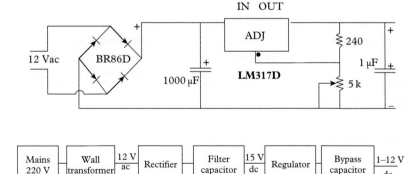

Figure B.1 *Circuit diagram for low-voltage power supply.*

Figure B.2 *Units of the LPVS.*

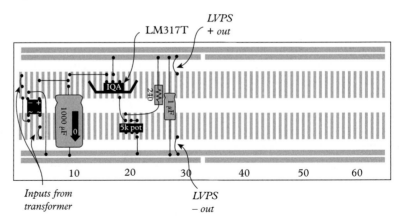

Figure B.3 *Spatial layout of the LPVS on a protoboard.*

The LM317T three-terminal integrated circuit (IC) contains 26 transistors. It keeps the output voltage constant with respect to an internal reference voltage using a feedback scheme. The regulator will hold its output voltage constant to within 0.5%. Consult Wikipedia for details of its working. It also protects itself against overload (too much current). Because it becomes hot, it is mounted on a "heat sink" having cooling fins, which help heat dissipation.

A resistor network, consisting of one variable resistance (a 5 kΩ potentiometer or pot) and one fixed one (240 Ω, 0.25 W) serves to adjust the output voltage. The pot, here used as a variable resistance, has the slider and one end connected, to ensure that some part of the pot resistance will be in the circuit, even if there is an uncertain contact inside the pot.

Finally, a 1 µF capacitor across the output bypasses high-frequency transients from either direction, either from the load or from the AC supply line. A convenient layout of the components on protoboard is shown in Fig. B.3.

High-voltage power supply (HVPS)

The high-voltage power supply will provide voltages from 100 V to 1000 V at a safe current of less than 1 mA. It is based on a power transistor driving a step-up transformer with feedback, so as to oscillate at around 200 kHz. About 50 to 500 V are generated and rectified by a half-wave voltage doubler to give the DC outputs. The device can be used for many electrostatic experiments, when combined with the electrometer described below, as fully described in [1].

The following components are needed:

- An LVPS, built according to the circuit described above
- 5 mH ferrite core inductor, on which are wound eight turns of wire, making an 8-to-500 transformer operating at approximately 200 kHz
- 2N3055 transistor and heat sink

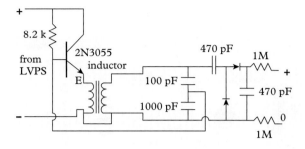

Figure B.4 *Circuit diagram for HVPS.*

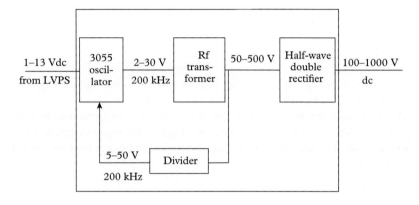

Figure B.5 *Block diagram showing voltage changes in the HVPS.*

- Resistors: 8.2 KΩ, 2 × 1 MΩ (0.5 W)
- Diodes: 2 × IN4007 (1 A, 1000 V)
- Ceramic disk capacitors: 100 pF, 1 kV; 1000 pF, 500 V; 2 × 470 pF, 1 kV
- Copper wire, enamel insulation, 26 gauge, 0.5 m.

The components are connected according to the circuit diagram in Fig. B.4.

The block diagram, shown in Fig. B.5, shows the various voltage changes in the circuit.

The 2N3055T transistor acts as a current amplifier: a small current to the base causes a much larger current to flow from collector to emitter. This large current flows through the eight turns of the primary of the transformer, creating the magnetic fields that induce much larger voltages in the 500 turns of the secondary, as in any transformer.

The voltage is divided down by a capacitative voltage divider (the 100 and 1000 pV capacitors), analogous to a resistive voltage divider. Some voltage is fed back to the transistor base, reinforcing the change that produced it. This is positive feedback and it causes the circuit to oscillate at about 200 kHz.

The two 470 pF capacitors and the two diodes join to make a voltage doubler. Into this comes alternating current up to 500 V: its output is DC at twice the input voltage. The two 1 MΩ resistors limit the current to safe values below 1 mA.

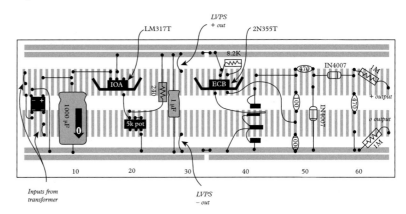

Figure B.6 *Spatial layout of the HPVS on a protoboard.*

More explanations of the circuits may be found in electronics books such as [2].

Notice that if you short the power supply, the resistors might have to dissipate $V^2/R = (1000)^2/2 \times 10^6 = 0.5$ W; hence the need for the larger-wattage resistors in this circuit. A convenient layout of the components on protoboard is shown in Fig. B.6.

Electrometer

We describe here an electrometer which, when used in tandem with the HVPS, can be used to conduct a series of experiments in electrostatics to detect and measure charges and their fields. Its heart is the operational amplifier OPA128, whose technical specifications and shielding needs can be found on the web. Components needed:

- The LVPS on its breadboard
- OPA128 op-amp with socket
- Capacitor: polyester, 0.01 μF, 100 V
- Resistors: $2 \times 110\Omega, 0.5$W; $1 \times 510\ \Omega$; 2×9.1 kΩ; 15 kΩ, 100 kΩ pot
- Two 4 in pie tins, for shielding.

The components are connected according to the circuit diagram shown in Fig. B.7.

O_1 is the least sensitive output reading when the multimeter is set to the 5 V range, but most sensitive with the meter on the 250 mV range. O_2 is the medium sensitive output readings with an AMM (20 kΩ/V) on the 250 mV range.

A convenient layout of the components on protoboard is shown in Fig. B.8: the positioning of the parts is important. Careful precautions must be taken to minimize the influence of static charges. (Refer to manuals on the web for the OPA128.) Note also that the input lead 3 is bent out from the IC and the 0.01 μF capacitor

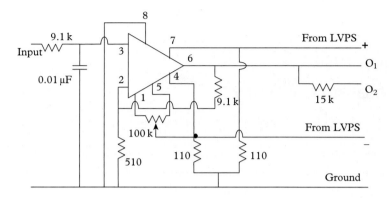

Figure B.7 *Circuit diagram for the electrometer.*

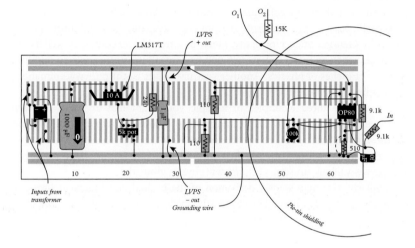

Figure B.8 *Spatial layout for building the electrometer.*

and 9.1 kΩ resistor are soldered to it to minimize leakage in socket and proto-board. The gain is ~19, but bypassing the 510 Ω with 100 Ω and putting 1 MΩ across the input turns the electrometer into an amplifier with gain of ~109 that is useful with small, low frequency signals.

....

REFERENCES

1. J.G. King and A.P. French, *Experiment Instructions for Physics 8.01* (MIT, Cambridge, MA, 1998).
2. P. Horowitz and W. Hill, *The Art of Electronics* (Cambridge University Press, Cambridge, 1990).

Appendix C
Reference library

In Wikipedia you will find useful articles about many of the topics in the projects. In addition, a search engine on the web will direct you to innumerable sites, maintained by universities and colleges all over the world, providing tutorial introductions to most of the topics mentioned in this book. Explore them and use them, in combination with texts. An excellent site to start with is HyperPhysics, <http://hyperphysics.phys-astr.gsu.edu/hbase/hph.html>, maintained by Georgia State University. Here, in a tree-like structure, as well as in a useful alphabetical list of topics, you will find introductory tutorials on and links to many branches of physics. Another very useful site is <http://www.animations.physics.unsw.edu.au>.

Let us give three examples of a topic you might like to revise or learn. Suppose you would like to see a derivation of Fresnel equations, which is not on the HyperPhysics site. A search engine with the key words "Fresnel equations derivation" will land you on three useful sites: <http://optics.hanyang.ac.kr>, <http://www.teknik.uu.se>, <http://www.physics.gmu.edu>.

Similarly, you may want a derivation of the cycloid and brachistochrone, in connection with Chapter 6. The key words "cycloid motion derivation" will direct you to Scott Morrison's tutorial at <http://tqft.net/papers/cycloid>.

In connection with a project in Part 3 on acoustics, how is acoustic impedance defined? A quick answer will be obtained via the key words "acoustic impedance definition derivation," which will direct you to <http://www.animations.physics.unsw.edu.au>, maintained by the University of New South Wales. A more thorough discussion and examples will be gained by reading one of the books on acoustics mentioned below.

The following books are useful. For in-depth study of a particular subject, peer-reviewed textbooks and monographs are reliable and indispensable. You can expand the list as new books are published, as the need arises, and as personal taste for texts develop, as readers will strike out on their own and increase the variety and scope of projects carried out. Find out which of the books in this list is available as a free download on the site <http://ebookbrowse.com>.

General

Y. Beers, *Introduction to the Theory of Errors* (Addison-Wesley, Cambridge, 1953).
J.R. Taylor, *An Introduction to Error Analysis* (University Science Books, Sausalito, 1997).

E.B. Wilson, *An Introduction to Scientific Research* (Dover, New York, 1990).
S.B. Parker, *Acoustics Source Book*, McGraw-Hill Science and Reference Series (McGraw-Hill, New York, 1988).
S.B. Parker, *Fluid Mechanics Source Book*, McGraw-Hill Science and Reference Series (McGraw-Hill, New York, 1988).
S.B. Parker, *Optics Source Book*, McGraw-Hill Science and Reference Series (McGraw-Hill, New York, 1988).
W.M. Haynes (ed.), *Handbook of Chemistry and Physics* (CRC, Boca Raton, FL, 2000), or any later edition.

Acoustics

A.H. Benade, *Fundamentals of Musical Acoustics* (Dover, New York, 1990).
P.M. Morse and K.U. Ingard, *Theoretical Acoustics* (McGraw-Hill, New York, 1954).
L.E. Kinsler and A.R. Frey, *Fundamentals of Acoustics* (Wiley, New York, 1962), or later editions to 2002.
T.D. Rossing, F.R. Moore, and P.A. Wheeler, *The Science of Sound* (Addison-Wesley, Reading, 2002).

Electromagnetism

D. Griffiths, *Introduction to Electrodynamics* (Prentice Hall, Upper Saddle River, NJ, 1998).
E.M. Purcell, *Berkeley Physics Course*, vol. 2 (McGraw-Hill, New York, 1986).

Electronics and communication

W. Faissler, *An Introduction to Modern Electronics* (Wiley, New York, 1991).
P. Horowitz and W. Hill, *The Art of Electronics* (Cambridge University Press, Cambridge, 1990).
F.M. Mimms, *The Forrest Mimms Circuit Scrapbook*, 2 volumes (LLH Technology Publishing, Eagle Rock, VA, 2000).
M.J. Wilson and S.R. Ford (eds), *The ARRL Handbook for Radio Communications* (ARRL, Newington, CT, 2009).

Experiments and projects

T.B. Brown, *The Taylor Manual of Experiments* (Addison-Wesley, Reading, 1959).
N.A. Downie, *Vacuum Bazookas, Electric Rainbow Jelly* (Princeton University Press, Princeton, 2001).

N.A. Downie, *Ink Sandwiches, Electric Worms* (The Johns Hopkins University Press, Baltimore, 2003).
C. Isenberg and S. Chomet, *Physics Experiments and Projects*, vols. I and II (Hemisphere and Taylor Francis, London, 1985, 1989).
J. Walker, *The Flying Circus of Physics* (Wiley, New York, 2006).
International Young Physicist Tournament, online at <http:/www.iypt.org>.
Physics Education, 2007, **1**.

Heat and thermodynamics

C.J. Adkins, *Equilibrium Thermodynamics* (Cambridge University Press, Cambridge, 1983).
H.S. Carlslaw and J.C. Jaeger, *Conduction of Heat in Solids* (Oxford University Press, Oxford, 1968).
F. Reif, *Fundamentals of Statistical and Thermal Physics* (McGraw-Hill, New York, 1965).

Liquids

P.A. Egelstaff, *An Introduction to the Liquid State* (Oxford University Press, Oxford, 1994).
T.E. Faber, *Fluid Dynamics for Physicists* (Cambridge University Press, Cambridge, 1995).
D. Tabor, *Gases, Liquids and Solids* (Cambridge University Press, Cambridge, 1991).
A.J. Walton, *Three Phases of Matter* (Oxford University Press, Oxford, 1983).
J-P. Hansen and I.R. McDonald, *Theory of Simple Liquids* (Academic Press, New York, 2013).

Mechanics

A.P. French, *Newtonian Mechanics* (W.W. Norton, New York, 1971).
H. Goldstein, *Classical Mechanics* (Addison-Wesley, Reading, 1980), and later editions.
R.C. Hilborn, *Chaos and Nonlinear Dynamics* (Oxford University Press, New York, 2001).
J.M. Knudsen and P.G. Hjorth, *Elements of Newtonian Mechanics* (Springer, Berlin, 1996).
Y.C. Fung, *A First Course on Continuum Mechanics* (Prentice Hall, Upper Saddle River, NJ, 1993).

Optics

R.D. Guenther, *Modern Optics* (Wiley, New York, 1990).
E. Hecht, M. Coffey, and P. Dolen, *Optics*, 4th ed. (Addison-Wesley, Reading, 2002).
S.G. Lipson, H. Lipson, and D.S. Tannhauser, *Optical Physics* (Cambridge University Press, Cambridge, 1995).
E.G. Steward, *Fourier Optics* (Ellis Horwood, Chichester, 1983).

Solid state physics

C. Kittel, *Introduction to Solid State Physics* (Wiley, New York, 1995), or later editions.
G. Strobl, *The Physics of Polymers* (Springer, Berlin, 2007).
S.M. Sze, *Semiconductor Sensors* (Wiley-Interscience, New York, 2005).

Sport and flight

J. Anderson, *Introduction to Flight* (McGraw-Hill, New York, 2007).
D.I. Anderson and S. Eberhardt, *Understanding Flight* (McGraw-Hill, New York, 2009).
A. Armenti, *The Physics of Sports* (American Institute of Physics Press, Melville, NY, 1996).

Waves

F.S. Crawford, *Berkeley Physics Course*, vol. 3 (McGraw-Hill, New York, 1968).
A.P. French, *Vibration and Waves* (W.W. Norton, New York, 1971).
I.G. Main, *Vibration and Waves in Physics* (Cambridge University Press, Cambridge, 1994).

Index

A
acoustic radiation resistance, 164
acoustics
    cavity resonators and filters, 138–40
    loudspeakers, 124–8
    musical instruments, 146–59
    resonance in pipes, 134–7
    room acoustics, 141–5
    rubber bands and cords, 45–6, 48
    sound from boiling kettle, 288–9
    sound from gas bubbles in liquid, 163–7
    tuning forks, 129–33
    vibrating wires and strings, 117–23
actuators, 295
aerosols, light scattering, 282–3
Alberti's windows, 53
amplifiers, 294–5
    two-stage, 103
analog multimeters, 298–9
analog oscilloscopes, 300–1
Antoine equation, 266
Arrhenius relation, 174

B
balls, bouncing, 13–16
band pass filters, 296
bandwidth limiting filter, 99
barrier penetration, 212–14
beaded strings, 117–18, 119, 120–1
bent tuning curve, 277–8
bifilar suspension, 50–1, 53
birefringence
    cellulose tape, 207–11
    rubber bands, 46
blowout, lamp, 304
books, useful, 310–13
bottle emptying, 197–200
bouncing balls, 13–16
brachistochrone, 38, 41
brass musical instruments, 155–9
Brewster angle, 221–3
Brownian motion, 228
bubbles
    shape and path of air bubbles in liquid, 168–71
    sound from gas bubbles in liquid, 163–7
Bunsen oil spot photometer, 99–100
Burgers vortex, 193
burnout lifetime, lamp, 57–8, 60

C
camera shake, 231
candles
    flame properties, 286–7
    oscillation (burning at both ends), 285
capillary waves, 273–4
castor oil, refractive index, 82
cavity resonators, 138–40
cellulose tape, birefringence in, 207–11
chain rotation, 34–7
characteristic impedance, 108
circular hydraulic jump, 188–91
Clausius–Clapeyron equation, 261, 281
clock pendulum, 40
coefficient of restitution, 14
coherence length, 96, 101
compass needle, non-linear dynamics, 276–7
conduction time, 246–7
connectors, 299–300
contact angle, surface roughness and, 281–2
cooling, 245–53, 287
corrugated plastic pipe (whirly), 288
coupled Euler–Lagrange equations, 51
cycloid paths, 38–43

D
Dalton's law of partial pressures, 268
Darcy's law, 281
DC motors, 295
Debye–Sears method, 185
diatomic lattice, 107–8
dielectrics, microwaves in, 80–8
differential light scattering, 232–5
diffraction
    from a sound source, 126–7
    of light by ultrasonic waves, 184–7
    spots, 181, 183
    surface ripples, 180–3
diffusion, ink in water, 173–5
digital multimeters, 297–8
digital oscilloscopes, 301
dipole interactions, 69–73
dispersion relationship, 106
dominoes, falling, 25–7
Doppler effect, 89–93
drinking bird, 260–5
drop formation in falling stream of liquid, 273–4
drop oscillations, 258–9
Duffing equation, 277–8

E
electrical instruments, 297–302
electrolytes, microwave transmission, 83–4
electromagnetism
    Doppler effect, 89–93
    Elihu Thomson jumping ring, 74–9
    incandescent lamps, 57–61
    Johnson noise, 95, 102–5
    jumping ring, 74–9
    magnetic dipoles, 69–73
    microwaves in dielectrics, 80–8
    network analogue for lattice dynamics, 106–10
    noise analysis, 94–101
    resistance networks, 111–14
    solenoid propulsion, 62–8
electrometer, 308–9
Elihu Thomson jumping ring, 74–9
equilibrium configuration, slinky, 18, 21
evanescent wave, 212

F
fan rotation speed, 91
Faraday instability (waves), 278
Faraday law, 70
Fermat's minimum time principle, 38
FFT program, 96
Fick's law, 262
film boiling, 254
filters
    acoustic, 138–40
    band-pass, 296
    bandwidth limiting, 99
    circuits, 100
    Lyot, 208, 210–11
    noise-reducing, 296
flag flapping, 283–4
flame properties, 286–7
"forbidden gap", evanescent wave, 212
Fourier transform lens, 230

Fresnel diffraction zones, 127–8
Froude number, 193
fundamental resonance frequency, 124–5, 127

G

gas discharge from pressurized tanks, 285
granular materials, 279–80
gravitational surface waves in water, 274–5
grid of resistances, 112–13
guitar acoustics, 118, 119, 122, 151–4

H

haze testing, 282–3
heat, see temperature and heat,
Helmholtz coil, 71, 276
Helmholtz resonators, 138–40
hexagonal pencil rolling, 279
high-voltage power supply, 306–8
Hooke's law, 44
hot spot, 58
Huygen's pendulum, 40, 43
hydraulic jump, 188–91

I

incandescent lamps, 57–61
ink diffusion in water, 173–5

J

Johnson noise, 95, 102–5
jumping ring of Elihu Thomson, 74–9

K

kettle boiling, sound from, 288–9

L

ladder of resistances, 111–12
laser Doppler interferometry, 92–3
laser speckle, 226–31
lateral quadrupole, 129–30
lattice dynamics, network analogue, 106–10
Leidenfrost effect, 254–9

light
  birefringence in cellulose tape, 207–11
  differential scattering, 232–5
  diffraction by ultrasonic waves, 184–7
  intensity from a line source, 236–8
  interference in reflecting tubes, 239–42
  polarization by transmission, 221–5
  reflection and transmission, 215–20
  scattering by aerosols, 282–3
  scattering by surface ripples, 180–3
  scattering from suspensions, 232–5
  transmission coefficient, 212, 213
  transmission in rubber bands, 46, 47
  transmission through stacked dielectric plates, 84–5
light bulbs, 57–61
linear quadrupole, 130
liquids
  circular hydraulic jump, 188–91
  diffraction of light by ultrasonic waves, 184–7
  drop formation and capillary waves in falling stream of liquid, 273–4
  gravitational surface waves in water, 274–5
  ink diffusion in water, 173–5
  light scattering by surface ripples, 180–3
  liquid–vapor equilibrium, 266–9
  plastic bottle oscillator, 197–200
  refractive index gradients, 176–9

  salt water oscillator, 201–4
  shape and path of air bubbles, 168–71
  sound from gas bubbles, 163–7
  surface wave patterns in vertically oscillating liquid, 278–9
  vortex physics, 192–6
longitudinal pulse velocity, 17–18, 23
lossless media, microwave transmission, 82–3
loudspeakers, 91–2, 124–8
low-pass filter and rectifier, 99
low-voltage power supply, 305–6
Lyot filter, 208, 210–11

M

magnetic dipoles, 69–73
magnetic force law, 64–5
magnetostatic force, 71, 72
Mathieu equation, 274
mechanical instruments, 297
mechanics
  bouncing balls, 13–16
  cycloid paths, 38–43
  oscillation modes of a rod, 49–53
  pulse speed of falling dominoes, 25–7
  rotating vertical chains, 34–7
  rubber bands and cords, 44–8
  soft springs (slinky), 17–24
  variable mass oscillator, 28–33
metronome synchronization, 275–6
microwaves in dielectrics, 80–8
Mie scattering, 282
Moka espresso coffee maker, 280–1
Moody charts, 285
multimeters, 297–9
musical instruments
  brass, 155–9
  guitar, 118, 119, 122, 151–4

  piano, 118–19, 120, 122
  violin, 146–50

N

network analogue for lattice dynamics, 106–10
Newton's law of cooling, 245
noise analysis, 94–101
  $1/f$ noise, 95
  Johnson noise, 95, 102–5
noise generator circuit, 98
noise-reducing filters, 296
non-linear dynamics, compass needle, 276–7
nucleate boiling, 254
Nyquist theorem, 102

O

optics
  barrier penetration, 212–14
  birefringence in cellulose tape, 207–11
  birefringence in rubber bands, 46
  instruments, 302–3
  laser speckle, 226–31
  light intensity from a line source, 236–8
  light interference in reflecting tubes, 239–42
  light scattering from suspensions, 232–5
  polarization by transmission, 221–5
  reflection and transmission of light, 215–20
  see also light
oscillation/oscillators
  drop oscillations, 258–9
  oscillating candle, 285
  oscillation modes of a rod, 49–53
  plastic bottle oscillator, 197–200
  salt water oscillator, 201–4
  variable mass oscillator, 28–33
oscilloscopes, 300–1, 303–4
osmotic pressure measurement, 177, 178

P

paper, mechanical properties, 284
particle image velocimetry, 192, 194
partner choice, 7
pencil rolling, 279
pendulum
  chaotic, 52
  double, 51, 53
  horizontal, 50, 52
  Huygen's, 40, 43
  torsional, 52
  vertical, 49–50, 52
photography
  camera shake, 231
  laser speckle, 229–31
piano strings, 118–19, 120, 122
pink noise, 95
pipes
  acoustic resonance, 134–7
  corrugated plastic (whirly), 288
plastic bottle oscillator, 197–200
polarization
  rubber bands, 46
  by transmission, 221–5
popping frequency, mouthpiece, 156
position sensor, 296
power supplies, 294
  circuit diagrams, 304–9
Poynting vector time average, 236
prisms, 212–14, 215–16, 217–18
project, defined, 2–4
project lab
  devices, 294–6
  electrical instruments, 297–302
  facilities, 290–4
  heat and temperature instruments, 303
  materials and storage, 293–4
  mechanical instruments, 297
  optical instruments, 302–3
  tools, 290, 291
  workshop, 291
propulsion, solenoids, 62–8
pulse speed
  dominoes, 25–7
  slinky, 17–18, 23

Q

quadrupole radiation, 129–30

R

radiation field, 125–6, 127, 129–30, 131–2
radiation time, 245–6
Rankine vortex, 193
reference library, 310–13
reflecting tubes, light interference, 239–42
refractive index
  castor oil, 82
  gradients, 176–9
resistance networks, 111–14
Reynolds number, 193
ripple patterns, light scattering, 180–3
room acoustics, 141–5
rotating chains, 34–7
rotating fan speed, 91
rubber bands and cords, 44–8

S

salt water oscillator, 201–4
saturation temperature, 254
Scotch tape (sellotape), birefringence in, 207–11
sensors, 295–6, 303–4
signal generators, 301–2
slinky, 17–24
Snell's law, 176
soils, water content, 86
solar radiation flux, 270–2
soldering, 291–2
solenoid propulsion, 62–8
speckle photography, 229–31
spectral analysis
  analog methods, 95–6
  digital methods, 96
spherical flasks, light transmission, 216–17, 218, 219–20
springs, soft (slinky), 17–24
standing waves, 19
Stefan–Boltzmann law, 248
Stokes–Einstein relation, 175
stress–strain curves, 45, 46–7
surface tension, 181, 273
synchronized motion, 275–6

T

tank pressure, 285
tautochrone property, 39
temperature and heat
  cooling, 245–53, 287
  drinking bird, 260–5
  drop oscillations, 258–9
  instruments, 303
  Leidenfrost effect, 254–9
  liquid–vapor equilibrium, 266–9
  rubber bands, 46, 47–8
  solar radiation flux, 270–2
thermal (Johnson) noise, 95, 102–5
thermal radiation resistance, 164
thermopile, 59
timber, absorption and defects, 87
tools, 290, 291
total internal reflection, 212
transformation mode, 1
transformers, 294
transition boiling stage, 254
transmission coefficient, 212, 213
transmission mode, learning in, 1
tubes, light interference, 239–42
tungsten filament lifetime, 57–9
tuning curves, bent, 277–8
tuning forks, 129–33
two-dimensional grid of resistances, 112–13
two-stage amplifier, 103

U

ultrasonic waves, diffraction of light in liquids, 184–7

V

variable mass oscillator, 28–33
velocity sensor, 296
vibrating wires and strings, 117–23
violin acoustics, 146–50
vortex physics, 192–6

W

water content of soils, 86
whirly toy, 288
wood, absorption and defects, 87
workshop, 291

Z

Zener diode, 96